Memoirs of the American Mathematical Society

Number 174

Jorge Ize

Bifurcation theory for Fredholm operators

Published by the

AMERICAN MATHEMATICAL SOCIETY

Providence, Rhode Island

VOLUME 7 · NUMBER 174 (first of 4 numbers) · SEPTEMBER 1976

ABSTRACT

This work is concerned with the study of bifurcation for the equation $Ax-T(\lambda)x = g(x,\lambda)$ where A is a Fredholm operator of non-negative index between two real or complex Banach spaces, $T(\lambda)$ is a continuous operator analytic in λ and $g(x,\lambda)$ satisfies the usual smallness conditions. The notion of algebraic multiplicity is defined for $A-T(\lambda)$ (which coincides with the usual definition in case $T(\lambda)$ is λI). Under the condition that the codimension of the range of $A-T(\lambda)$ is zero for small nonzero λ and using, in the real case, degree theory and in the complex case the suspension of the Hopf map, it is proved that if $i(A)$, the index of A, is zero and the multiplicity is odd then bifurcation occurs while if $i(A)$ is positive there is an $i(A)$-parameter family of bifurcating solutions.

Then higher homotopy groups of spheres are used to study the same bifurcation problem for the case of several parameters $\lambda_1,\ldots,\lambda_n$.

If A is the identity, $T(\lambda)$ and $g(x,\lambda)$ are compact and λ_o is a bifurcation point it is proved that the connected component of $(0,\lambda_o)$ of the closure of the nontrivial solutions, if bounded, contains an even number of characteristic values of odd algebraic multiplicity. This is done by introducing stable cohomotopy. An application to global bifurcation of periodic orbits for dynamical systems is given.

Finally the case where $g(x,\lambda)$ has the form $N(x,\lambda)+S(x,\lambda)$ with $N(x,0)$ a homogeneous operator and S a small perturbation is treated and sufficient conditions for bifurcation along an isolated eigenray are derived.

A M S (M O S) 1970 Subject Classifications: Primary: 47H15, 35G20. Secondary: 55C25, 55E55, 47B30, 34C99, 45G99.

KEY WORDS AND PHRASES: Nonlinear functional analysis, bifurcation, Fredholm Operator, algebraic multiplicity, degree, Hopf map, Homotopy groups of spheres, global solutions, stable cohomotopy theory, eigenray, gradient operator.

Library of Congress Cataloging in Publication Data **CIP**

Ize, Jorge, 1946-
Bifurcation theory for Fredholm operators.

(Memoirs of the American Mathematical Society ; no. 174)
Bibliography: p.
1. Fredholm operators. 2. Differential equations, Partial--Numerical solutions. 3. Differential equations, Nonlinear--Numerical solutions. 4. Homotopy groups. I. Title. II. Series: American Mathematical Society. Memoirs ; no. 174.
QA3.A57 no. 174 [QA329.2] 510'.8s [514'.24]
ISBN 0-8218-2174-1 76-25186

TABLE OF CONTENTS

	page
INTRODUCTION	v
ACKNOWLEDGEMENTS	viii
CHAPTER ONE: BIFURCATION THEORY FOR ONE PARAMETER	1
§ I. Notations and preliminaries	2
§ II. Topological tools	6
§III. Finite multiplicity	12
§ IV. Bifurcation for Fredholm operators of non-negative index	24
§ V. Isolated eigenvalue versus finite multiplicity	35
§ VI. Bifurcation theory for $A(x) - T(\lambda)x = g(x,\lambda)$	42
CHAPTER TWO: BIFURCATION THEORY FOR SEVERAL PARAMETERS	48
§ I. A special case $d = d^* \leq n$	51
§ II. General situation	56
§III. Application: The Hopf bifurcation theorem	61
CHAPTER THREE: SOLUTIONS IN THE LARGE	70
§ I. Infinite dimensional cohomology	72
§ II. Global solutions	77
§III. The Hopf bifurcation theorem: global version	93
CHAPTER FOUR: SPECIAL NONLINEARITIES	98
§ I. Study of eigenrays	101
§ II. Application: Bifurcation near an isolated eigenray	105
APPENDIX	118
BIBLIOGRAPHY	125

INTRODUCTION

Bifurcation theory is the study of small solutions of non-linear equations, depending on a number of parameters, when for certain values of these parameters more than one solution may appear.

The scope of applications of this theory is thus extremely large, ranging from Biology to Physics, using tools as varied as functional analysis, topology, partial differential equations or numerical analysis.

Because of the difficulties encountered in obtaining explicit solutions to such problems, one tries to approximate the initial equation by a simpler one and from the properties of the solutions to the latter gain some insight for those of the former. If the nonlinear equation is a smooth function of its variables, a standard approximation (when justified, see e.g. [22]) is to linearize the equation.

Among the techniques used to study such problems, the topological method is particularly attractive because it gives quick qualitative results, such as existence of a solution, even if for computational purposes their usefulness is rather limited. Many successful applications of topology have been made in this context, some dating back to the beginning of this century with work by Poincaré, Schauder, Leray, Lusternick, Schnirelman, Morse and others, and some quite recent in the setting of global analysis and fixed point theory.

The object of this paper is to try to give a new look at one of the most basic ideas in this topological approach, that is deformation of maps or homotopy theory, and by doing so include the well known Leray-Schauder degree theory in the setting of generalized degree theories.

For this reason, the general philisophy of this paper is to give simple, direct, proofs of some very often used results as well as new ones, in the spirit of [2], [22], [26], [38]. So, except for the Hopf bifurcation problem, almost no example of applications are included here and one is referred to the very extensive literature on the subject, e.g. [5], [6], [27], [30], to name a few.

Chapter one is concerned with the study of bifurcation for equations with only one parameter, real or complex, entering. The type of equation studied is of the form: $Ax - T(\lambda)x = g(x,\lambda)$ where A is a Fredholm operator of non-negative index, $T(\lambda)$ is a continuous linear operator analytic in λ and $g(x,\lambda)$ is a small non-linearity. In section I, the basic definitions and assumptions are made; section II considers the main topological tools used in the chapter, i.e. degree theory and the Hopf map. The homotopy class of a particular map is computed. Section III is concerned with the special case of an eigenvalue of A which has finite multiplicity and which gives a splitting of the Banach space considered. In section IV, the main result for this type of bifurcation is proved, that is, after a reduction à la Liapounov-Schmidt, natural extensions of Krasnosel'skii theorem, [22], are given for the real and complex cases and operators of index zero. Also the existence of a family of solutions for the positive index case is established. Section V is rather technical in nature but enables one to see the relationship between eigenvalues which are isolated and eigenvalues which have finite multiplicity. Lastly, section VI extends all the previous results to the case where $T(\lambda)$ replaces λI.

The setting of section VI leads naturally to chapter two and the study of bifurcation theory when more than one parameter are involved. In this case, it is impossible to give simple criteria to prove the existence of a solution, and the attempt is to give a philosophical viewpoint for the study of such problems. So, section I is concerned with the case where there are enough parameters in the bifurcation equation to use degree theory, section II gives some sort of recipe to study the general case, and section III applies these ideas to the bifurcation of periodic orbits for autonomous system.

Chapter three goes from the local problem to study what happens to a bifurcating solution in the large. Thus a slight sharpening of the important result of Rabinowitz' global theorem [26] is given for the real case and, in the complex case, the notion of stable cohomotopy and its characterization due to Geba [13] and Granas [17] is used to prove a similar result. Finally the global version of the Hopf bifurcation problem, due to Alexander and Yorke, is treated.

At last, chapter four returns to the local problem, but with new conditions on the nonlinearity. Hence, generalizations of results of Sather [30] and Kirchgässner [21] are proved. Here again no attempt at the greatest generality has been made, but rather a method and a complement to Dancer's articles [9] and [10] are given.

ACKNOWLEDGEMENTS

This paper is an extended version of the author's doctoral thesis written at the Courant Institute of New York University. The author would like to express his gratitude to Professor Louis Nirenberg for his advice, for his insight and his kindness. Thanks are also due to the Consejo Nacional de Ciencia y Tecnología of Mexico which provided a partial support for the author's doctoral studies and in particular for this research.

BIFURCATION THEORY FOR FREDHOLM OPERATORS

by JORGE IZE

CHAPTER ONE

BIFURCATION THEORY FOR ONE PARAMETER

Problems of bifurcation arise in the study of equations of the form:

$$(1) \qquad F(x,\lambda) = 0$$

where F is a smooth function of x, which belongs to some normed space X, and of the one dimensional real or complex parameter λ. It is usually assumed that $F(x,\lambda)$ has a known family of solutions, say $F(0,\lambda) \equiv 0$, referred to as the <u>trivial solutions</u>. One is then interested in determining the existence of non-trivial solutions of (1), i.e. with x not zero. This leads to the following:

<u>Definition</u>: $(0,0)$ is a <u>bifurcation point</u> of (1) with respect to the line of trivial solutions, if every neighbourhood of $(0,0)$ in the $X \times \lambda$ space contains non-trivial solutions.

The natural approach to the study of (1) is then to expand $F(x,\lambda)$ in a Taylor series at $(0,0)$, assuming F is smooth enough and defining such a series in the appropriate spaces. So

$$F(x,\lambda) = F_x(0,0)x + \lambda F_{x\lambda}(0,0)x + o(||x||,|\lambda|)$$

Received by the editors January 2, 1975
Research partially supported by the CONACYT of Mexico

If $F_x(0,0)$ is invertible, the adequate version of the implicit function theorem will give a unique solution of (1), and hence no bifurcation.

The next case to study is when $F_x(0,0)$ fails to be an invertible operator by a finite amount: this is the setting for Fredholm operators. A more important reason to study Fredholm operators in this context is that many concrete problems are of this form: For example a large class of elliptic differential operators, (Friedman [11], Goldberg [14]), integral operators, operators of the form I-C, where I is the identity and C is compact, etc....

The aim of this chapter is to study the equation:

$$Ax - T(\lambda)x = g(x,\lambda) \tag{2}$$

with $T(\lambda) = \lambda T_1 + \ldots + \lambda^p T_p + \ldots$, where A is a Fredholm operator of non-negative index, and g is "small" near (0,0). More specific hypothesis will be taken up in the next section.

§ I: NOTATION AND PRELIMINARIES

B and E will denote two Banach spaces over the field $\mathbb{K}$ of the real or complex numbers.

A is a linear operator defined on B with range in E.

<u>Definition 1.1</u>: A is a <u>Fredholm</u> operator if:

1) A is a <u>continuous</u> operator.

2) ker A is a finite dimensional subspace of B, with dimension denoted by n(A) or sometimes d.

3) the range of A, R(A), is a finite codimensional subspace of E, with codimension denoted by d(A) or also d*.

It is well known (e.g. Goldberg [14]) that $R(A)$ is a closed subspace of E.

Except in the last section of this chapter, B will be continuously embedded in E, and the inclusion operator is taken to have norm no greater than one.

In this context and if B is not specified, some authors prefer to use the following:

<u>Definition 1.2:</u> An operator A on E, with domain $D(A)$, is a <u>Fredholm</u> operator if:

1)' A is a <u>closed</u> operator

2)' $\ker A$ is a finite dimensional subspace of E

3)' $R(A)$ is a finite codimensional subspace of E

<u>Remark 1.3:</u> In fact this is not a generalization, since it is easily checked that if one gives to $D(A)$ the graph norm $||x||_{D(A)} = ||x||_E + ||Ax||_E$ then $D(A)$ is a Banach space, A is continuous with respect to this new norm and $D(A)$ is continuously embedded in E.

In the rest of the paper only Definition 1.1. will be used, although it is sometimes more convenient, when B is included in E, to work only with the E norm. This is possible according to the following:

<u>Lemma 1.4:</u> If B is continuously embedded in E, let A be a Fredholm operator from B into E. Then A is a closed operator in the E-norm. This implies that A is a Fredholm operator in the sense of Definition 1.2.

<u>Proof:</u> Denote B by $D(A)$

Let $x_n \in D(A)$ with

$$||x_n - x||_E \to 0 \quad \text{and} \quad ||Ax_n - y||_E \to 0$$

as n tends to ∞, for some x and y in E. One needs to prove that: $x \in D(A)$ and $Ax = y$.

Since $\ker A$ is finite dimensional, one can decompose B as: $B = \ker A \oplus B_2$, where B_2 is a closed subspace of B (in the B-norm). Hence x_n admits the unique decomposition: $x_n = u_n + v_n, u_n \in \ker A, v_n \in B_2$. Moreover A is a one to one mapping from B_2 into $R(A)$, which are both Banach spaces, so, from the open mapping theorem (Taylor [36]), A restricted to B_2 admits a continuous inverse K from $R(A)$ into B_2. Now $Av_n = Ax_n \overset{E}{\to} y$ and since $R(A)$ is closed: $y \in R(A)$ i.e. $y = Az$ for some z in B_2. Then by continuity of K: $v_n \overset{B_2}{\to} z$, in particular $||v_n||_B$ is bounded.

a) Suppose: $||x_n||_B \leq M$ for some M,

then u_n are bounded in the finite dimensional space $\ker A$. So there exists a subsequence u_{n_i} such that: $u_{n_i} \overset{B}{\to} u$ in $\ker A$.

So $x_{n_i} \overset{B}{\to} u + z$ and $||x_{n_i} - (u+z)||_E \leq ||x_{n_i} - (u+z)||_B \to 0$

One can then conclude that: $x = u + z$ belongs to $B = D(A)$ and that $Ax = Az = y$.

b) Suppose, on the contrary, that $||x_n||_B$ are not bounded. So there exists a subsequence x_{n_i} such that: $||x_{n_i}||_B \to \infty$. Now $v_{n_i} \overset{B}{\to} z$ so $\dfrac{v_{n_i}}{||x_{n_i}||_B} \overset{B}{\to} 0$ and:

(1.5) $$\frac{u_{n_i}}{||x_{n_i}||_B} = \frac{x_{n_i}}{||x_{n_i}||_B} - \frac{v_{n_i}}{||x_{n_i}||_B}$$ are also bounded.

Hence, as before, there is a subsequence, also denoted $\frac{u_{n_i}}{||x_{n_i}||_B}$, which converges in the B norm to some u in $\ker A$. From (1.5)

$$||\frac{x_{n_i}}{||x_{n_i}||_B} - u||_E \leq ||\frac{x_{n_i}}{||x_{n_i}||_B} - u||_B$$

$$\leq ||\frac{u_{n_i}}{||x_{n_i}||_B} - u||_B + ||\frac{v_{n_i}}{||x_{n_i}||_B}||_B \to 0$$

In particular: $||u||_B = 1$ and since $||x_{n_i} - x||_E \to 0$ then

$\frac{||x_{n_i}||_E}{||x_{n_i}||_B} \to 0$ so $u = 0$ which gives a contradiction with the fact that B is included in E.

Q.E.D.

The remaining hypothesis for equation (2) are listed below:

Except in the last section, $T(\lambda)$ will be of the form λI.

Finally $g(x,\lambda)$ is a continuous mapping: $B \times \mathbb{K} \to E$ and satisfies the condition, (H):

1) $g(0,0) = 0$

2) There are positive constants r_o, λ_o, C, such that for all r, x, x', λ, λ' with

$$||x||_B,\ ||x'||_B \leq r \leq r_o,\ |\lambda|, |\lambda'| \leq \lambda_o$$

then:

$$(1.6)\begin{cases} ||g(x,\lambda) - g(x',\lambda')||_E \leq C\left[M(r)||x-x'||_B + |\lambda-\lambda'|\right] \\ \\ ||g(x,\lambda)||_E \leq C\left[||x||_B^2 + |\lambda|^{2n+1}\right] \end{cases}$$

where $M(r)$ is a continuous function from $\mathbb{R}^+$ into $\mathbb{R}^+$, with $M(0) = 0$, and n is some large integer to be chosen later.

§ II : TOPOLOGICAL TOOLS

A) Essential maps

A precise study of equation (2) is in general not possible. However by means of deformation arguments, one can ascertain the existence of bifurcation under very general conditions. In this chapter a simple topological fact will be used.

Let T be a continuous map from B^n, a closed ball in $\mathbb{R}^n$, into $\mathbb{R}^k$. Assume that T_o, the restriction of T to the boundary S^{n-1} of B^n, maps S^{n-1} into $\mathbb{R}^k - \{0\}$.

Definition 2.1: T_o is said to be essential, with respect to B^n, if and only if every continuous extension $\tilde{T} : B^n \to \mathbb{R}^k$ of T_o has a zero.

The following classical observation, though trivial, is very useful. Its proof can be found in any book on homotopy theory.

Theorem 2.2: T_o is essential if and only if T_o is not homotopic to a constant map from S^{n-1} into $\mathbb{R}^k - \{0\}$, which represents the neutral element in $\pi_{n-1}(S^{k-1})$.

Since for $n < k$ any such map T_o is homotopically

trivial, only two cases are of interest:

a) $n = k$: it is then well known that T_o is essential if and only if degree $(T, B^n, 0) \neq 0$.

b) $n > k$: here one has to make use of higher homotopy groups of spheres and there is no simple characterization for essentiality. But of particular interest is the Hopf map, denoted η (Hopf [19]) from S^3 onto S^2.

Let S^3 be considered as the set: $\{(\lambda, z) \in \mathbb{C} \times \mathbb{C} \,/\, |\lambda|^2 + |z|^2 = 1\}$

η is the composition of:

$$S^3 \longrightarrow \mathbb{C}P(1) \xrightarrow{\sim} \mathbb{C} \cup \{\infty\} \xrightarrow{\sim} S^2$$

$$(\lambda, z) \longrightarrow [\lambda : z] \longrightarrow z/\lambda = w \longrightarrow \left(\frac{w + \bar{w}}{|w|^2 + 1}, \frac{w - \bar{w}}{i(|w|^2 + 1)}, \frac{|w|^2 - 1}{|w|^2 + 1} \right)$$

i.e.

$$(2.3) \qquad \eta : (\lambda, z) \longrightarrow (\mathrm{Re}\ 2\bar{\lambda} z, \mathrm{Im}\ 2\bar{\lambda} z, |z|^2 - |\lambda|^2)$$

This map plays a major role in the computation of the stable homotopy groups of spheres. Toda [35].

B) Suspension

The suspension is a well known topological construction (Spanier [34]). A simple version of it will be used here: Let ψ be a continuous map from S^{n-1} to S^{k-1}, two unit spheres. The suspension of ψ, $\Sigma\psi$, is a map from S^n to S^k, which is described geometrically as follows: Identify S^{n-1} with the equator of S^n and likewise for S^{k-1} in S^k. $\Sigma\psi$ is defined by mapping the north (south) pole of S^n to the north (south) pole of S^k and then by extending ψ to the hemispheres linearly on the great circles. This construction is expressed analytically as follows: Extend ψ to the unit ball B^n (e.g. by homogeneity of degree one) secondly define:

$$\Sigma\psi(x_1, \ldots, x_{n+1}) = \left(\frac{\psi(x_1,\ldots,x_n)}{(||\psi||^2+x_{n+1}^2)^{1/2}}, \frac{x_{n+1}}{(||\psi||^2+x_{n+1}^2)^{1/2}} \right)$$

It is the content of the Freudenthal suspension theorem (Spanier [34] p.458) that after a certain number of suspensions, the homotopy class of a map ψ does not change any more, it is then called the stable homotopy class of ψ. Precisely, $\Sigma^m\psi : S^{n-1+m} \to S^{k-1+m}$ is in the stable range for $m \geq n - 2k + 3$.

<u>Example 2.4:</u> $\pi_3(S^2)$ is a free abelian group generated by the Hopf map and for $n > 2$, $\pi_{n+1}(S^n)$ is isomorphic to a cyclic group of order two generated by the suspension of the Hopf map. (Toda [35] 5.3 and chapter XIV).

In this chapter, homotopy classes of maps F from S^{n-1} to $\mathbb{R}^k - \{0\}$ will be studied. It is then clear that, by considering $\psi(x) = \frac{F(x)}{||F(x)||}$, one can define the suspension of F and that, up to a positive factor which can be deformed to one, this suspension has the form:

(2.4) $$\Sigma F(x_1,\ldots,x_{n+1}) = (F(x_1,\ldots,x_n), x_{n+1})$$

C) <u>Application</u>

Of particular interest is the homotopy class of the map $F : \mathbb{K}^d \times \mathbb{K} \to \mathbb{K}^d \times \mathbb{R}$ defined on the sphere S_r:

$$\{(x,\lambda) \in \mathbb{K}^d \times \mathbb{K}, x = (x_1,\ldots,x_d) \ / \ ||x||^2 + |\lambda|^2 = r^2 + (Mr)^{2/n}\}$$

where r and Mr are small quantities and n is a large integer.

(2.5) $$F(x,\lambda) = (\lambda^{k_1} x_1,\ldots,\lambda^{k_d} x_d, \sum_1^d |x_i|^2 - r^2)$$

where $k_1,\ldots,k_d$ are arbitrary, but fixed, non negative integers.

Note that for convenience, $\mathbb{C}$ is identifie with $\mathbb{R}^2$ and so $\lambda^k x$ means $(\mathrm{Re}(\lambda^k x), \mathrm{Im}(\lambda^k x))$: so for $\mathbb{K} = \mathbb{R}$, (2.5) defines a map from S^d to $\mathbb{R}^{d+1}-\{0\}$ and, for $\mathbb{K} = \mathbb{C}$, a map from S^{2d+1} into $\mathbb{R}^{2d+1}-\{0\}$. The number $m = \Sigma_{i=1}^d k_i$ will be called the <u>algebraic multiplicity</u> of (2.5).

<u>Theorem 2.6:</u> If $\mathbb{K} = \mathbb{R}$; the degree of the mapping (2.5) is -2 (resp. 0) if m is odd (resp. even).

If $\mathbb{K} = \mathbb{C}$, the mapping (2.5) has the homotopy type of $\Sigma^{2d-2}(-m\eta)$.

Note that if $\mathbb{K} = \mathbb{C}$, the mapping (2.5) is <u>essential</u> if and only if: $d = 1$ and $m > 0$ or, $d > 1$ and m is odd. (See example 2.4)

<u>Proof:</u> The map (2.5) will be deformed to a simpler map:
On S_r one has: $||x||^2 - r^2 = ||x||^2(1-A) - |\lambda|^2 A$

with $A = (1 + (Mr)^{2/n} r^{-2})^{-1} < 1$.

So (2.5) is homotopic to $(\lambda^{k_1}x_1,\ldots,\lambda^{k_d}x_d, ||x||^2 - |\lambda|^2)$ via $(\lambda^{k_1}x_1,\ldots,\lambda^{k_d}x_d, ||x||^2(1-(1-t)A) - |\lambda|^2((1-t)A+t))$ for t in $[0,1]$.

It is easily checked that the latter deformation is valid. Then, if $x_i = z_i(r^2+(Mr)^{2/n})^{1/2}$ and $\lambda = \lambda_o(r^2+(Mr)^{2/n})^{1/2}$ one has

$||z||^2 + |\lambda_o|^2 = 1$ and (2.5) has the same homotopy type as:

$$(2.7) \qquad (\lambda^{k_1} z_1, \ldots, \lambda^{k_d} z_d, ||z||^2 - |\lambda|^2)$$

on a unit sphere. This map can be deformed to:

$$(2.8) \qquad (\lambda^m z_1, z_2, \ldots, z_d, |z_1|^2 - |\lambda|^2)$$

via a sequence of simple homotopies.

In fact (2.7) is homotopic to:

$$(\lambda^{k_1} z_1, \ldots, \lambda^{k_{d-2}} z_{d-2}, \lambda^{k_{d-1}+k_d} z_{d-1}, z_d, \Sigma_{i=1}^{d-1} |z_i|^2 - |\lambda|^2)$$

using the following deformation:

$$(\lambda^{k_1} z_1, \ldots, \lambda^{k_{d-2}} z_{d-2}, a_{11} z_{d-1} + a_{12} z_d, a_{21} z_{d-1} + a_{22} z_d, ||z||^2 - |\lambda|^2 - t|z_d|^2)$$

where $(a_{ij})_{i,j=1,2}$ represent the matrix:

$$\begin{bmatrix} 1-t & t \\ -t & 1-t \end{bmatrix} \begin{bmatrix} (1-t)\lambda^{k_{d-1}} & -t \\ t\lambda^{k_{d-1}+k_d} & (1-t)\lambda^{k_d} \end{bmatrix} \text{ with } t \in [0,1].$$

It is clear that this mapping is not zero on the unit sphere in the (z,λ) space.

By repeating this homotopy, one sees that (2.7) is deformable to (2.8).

<u>If $\mathbb{K} = \mathbb{R}$ and m is odd, i.e $m = 2n+1$.</u>

(2.8) is homotopic to $(\lambda z_1, z_2, \ldots, z_d, |z_1|^2 - |\lambda|^2)$ using the deformation $(t + (1-t)\lambda^{2n})\lambda z_1$. Then by solving $(\lambda z_1 = \varepsilon^2,$ $z_2 = 0, \ldots, z_d = 0, |z_1|^2 - |\lambda|^2 = 0)$ i.e. at $\lambda = z_1 = \pm\varepsilon, z_i = 0$

for $i = 2,\ldots,d$, the Jacobian of this map is of the form:

$$\begin{bmatrix} \lambda & 0 & \cdots\cdots & & z_1 \\ 0 & 1 & 0 & & \vdots \\ \cdot & 0 & 1 & & 0 \\ 0 & \cdot & 0 & 1 & \vdots \\ 2z_1 & 0 & \cdots\cdots & & -2\lambda \end{bmatrix}$$

with determinant $-2(\lambda^2+z_1^2) = -4\varepsilon^2$

hence the degree in this case is: -2.

Note that a change of the orientation in the $(z_1,\ldots,z_d,\lambda)$ space, may change the sign of the Jacobian. Here the natural orientation is used.

If $\mathbb{K} = \mathbb{R}$ and $m = 2n$, the same deformation gives $(z_1,\ldots,z_d, |z_1|^2-|\lambda|^2)$ deformable to $(0,0,\ldots 0,-1)$ via $((1-t)z_1,\ldots (1-t)z_d,(1-t)(z_1^2-\lambda^2)-t)$.

If $\mathbb{K} = \mathbb{C}$ one recognizes the $2d-2$ suspension of:

$$\phi(\lambda,z_1) = (\lambda^m z_1, |z_1|^2-|\lambda|^2) \text{ defined on } S^3.$$

Note that one has to perform a series of homotopies to bring (2.8) in the form (2.4) i.e. to $(\lambda^m z_1,|z_1|^2-|\lambda|^2,z_2,\ldots,z_d)$ this is done via:

$$(\lambda^m z_1,\ldots,\mathrm{Re}z_d,(1-t)\mathrm{Im}z_d-t(|z_1|^2-|\lambda|^2),t\ \mathrm{Im}z_d+(1-t)(|z_1|^2-|\lambda|^2))$$

and then

$(\lambda^m z_1, \ldots, (1-t)\mathrm{Re}z_d + t(|z_1|^2 - |\lambda|^2), t\,\mathrm{Re}z_d - (1-t)(|z_1|^2 - |\lambda|^2), \mathrm{Im}z_d)$.

ϕ is then the composition of: $S^3 \overset{\alpha}{\to} S^3 \overset{\eta}{\to} S^2$, where α: $(\lambda, z) \to (\lambda^m, z)$ is a map of degree m and induces a map α^*: $\pi_3(S^2) \to \pi_3(S^2)$ which maps η into $m\eta$: Cronin [6] and Spanier [34]. This finishes the proof of the theorem.

Remark 2.9: It might be of independent interest to give a proof of the fact: $\Sigma^2 - m\eta \sim \Sigma^2\eta$ if m is odd.
~ 0 if m is even.

Consider the following homotopies, defined from S^5 to $\mathbb{R}^5 - \{0\}$, where $m = 2n+\varepsilon, \varepsilon = 0,1$ according to the parity of m.

Deform first $(\lambda^m z_1, z_2, ||z||^2 - |\lambda|^2)$ to

$(\lambda^{n+\varepsilon} z_1, \lambda^n z_2, ||z||^2 - |\lambda|^2)$ using the inverse of the homotopy which gave (2.8) from (2.7). Then consider $((1-t)\lambda^{n+\varepsilon} z_1 + t\bar{\lambda}^{\varepsilon}\bar{z}_2, (1-t)\lambda^n z_2 - t\bar{z}_1, ||z||^2 - |\lambda|^2)$. Multiplying the first term by $\bar{z}_1$, the second by $\lambda^{\varepsilon}\bar{z}_2$ and adding these terms, it is easily seen that this homotopy is not zero on the unit sphere in the (z,λ) space.

If $\varepsilon = 0$, $(\bar{z}_2, -\bar{z}_1, ||z||^2 - |\lambda|^2)$ may be linearly deformed to $(0,0,-1)$ while if $\varepsilon = 1$, one can deform $(\lambda\bar{z}_2, -\bar{z}_1, ||z||^2 - |\lambda|^2)$ to $(\bar{\lambda} z_2, z_1, ||z||^2 - |\lambda|^2)$ using the following homotopy:

$$\begin{bmatrix} 1-2t & t(1-t) \\ -t(1-t) & 1-2t \end{bmatrix} \begin{bmatrix} \mathrm{Im}\,\lambda\bar{z}_2 \\ \mathrm{Re}-z_1 \end{bmatrix}$$

leaving all the other terms unchanged.
A linear transformation taking the (complex) vector (z_2, z_1) into (z_1, z_2) will give the result: a rotation will take (z_2, z_1) into $(-z_1, z_2)$ and a deformation of the form given above will change the sign of z_1.

§ III : FINITE MULTIPLICITY

In this section, the following assumptions are made:

1) B is continuously included in E

2) there exists a positive integer α such that:

a) $R(A^\alpha) = R(A^{\alpha+\ell})$ for all positive ℓ,

b) $\ker(A^\alpha) = \ker(A^{\alpha+\ell})$ for all positive ℓ.

c) $\dim \ker A^\alpha = m < \infty$.

d) $i(A) = 0$.

Note that $D(A^\alpha) = \{x \in E \mid x \in D(A) = B, Ax \in B, \ldots, A^{\alpha-1}x \in B\}$ and that it is not assumed that A^α has a zero index. This will be proved in section V of this chapter.

Lemma 3.1: If $B = \ker A^\alpha \oplus X_2$ then $E = A(X_2) \oplus \ker A^\alpha$.

Proof: 1) $R(A) = A(X_2) \oplus A(\ker A^\alpha)$.

For if $y \in A(X_2) \cap A(\ker A^\alpha)$ then $y = Ax_2 = Ax_1$.
So: $x_2 - x_1 \in \ker A \subseteq \ker A^\alpha$ which implies: $x_2 = 0$ and $y = 0$.

2) $A(X_2)$ is closed in E: Since X_2 is closed in B and A restricted to X_2 is a continuous operator with range $A(X_2)$ of finite codimension, this follows from [14] IV-1-13. In fact, since A $(\ker A^\alpha)$ has dimension $m - d$ (d dimension of $\ker A$) and $R(A)$ is of codimension d, $A(X_2)$ has codimension m.

3) $A(X_2) \cap \ker A^\alpha = \{0\}$.
Since if: $y = Ax_2$ and $A^\alpha y = 0$ then $A^{\alpha+1}x_2 = 0$.
But $\ker A^\alpha = \ker A^{\alpha+1}$ implies $A^\alpha x_2 = 0$.

That is : $x_2 \in X_2 \cap \ker A^\alpha = \{0\}$.

The lemma follows, since $A(\ker A^\alpha)$ is contained in $\ker A^\alpha$ which has the right dimension to complement $A(X_2)$.

Theorem 3.2: *Consider the equation*

(3) $Ax - \lambda x = g(x,\lambda)$

Suppose that $g(x,\lambda)$ *satisfies* (1.6) *with* n *replaced by* α. *Let* P *be the projection of* B *onto* ker A. *Then there exist two positive constants* M *and* r_o *such that:*

i) *for all* $r, r < r_o$, (3) *has a solution* (x,λ) *with* $||Px|| < r$, $|\lambda| = (Mr)^{1/\alpha}$. *This set of solutions will be called the set of trivial solutions.*

ii) *If* $m = \dim_{\mathbb{K}} \ker A^{\alpha}$ *is odd, for all* $r, r < r_o$, (3) *has a non-trivial solution with* $||Px|| = r$, $|\lambda| < (Mr)^{1/\alpha}$.

iii) *Moreover, if* $\mathbb{K} = \mathbb{C}$ and $\dim_{\mathbb{C}} \ker A = 1$, *then for any* m *the statement of* (ii) *holds.*

Remark 3.3: A simpler version of theorem 3.2, due to Krasnosel'skii [22], for operators of the form $I - \lambda_o K$, K a real compact operator, has been extensively used in the literature. Also, in the complex case, when the ascent α is one, Böhme, [4], was the first to use more sophisticated topological tools, namely the Adams theorem for linearly independent vector fields on a sphere.

Note also that theorem 3.4 will give a sharper result than (iii) which is included here only because it is a direct consequence of the method used.

Finally, in the complex case, if $g(x,\lambda)$ is complex analytic with $g(0,\lambda) = 0$, $||g(x,\lambda)|| = o(||x||)$, Dancer [10] has shown that (0,0) is always a bifurcation point.

Proof of Theorem 3.2.

a) The Liapounov-Schmidt reduction.

Any x in B can be written as:

$$x = x_1 + x_2$$

with x_1 in $\ker A^\alpha$, x_2 in X_2, and A restricted to X_2 (being continuous onto AX_2) has a continuous inverse K.

Then, from Lemma 3.1, (3) is equivalent to:

$$Ax_2 - \lambda Qx_2 = Qg(x_1+x_2,\lambda)$$

$$Ax_1 - \lambda x_1 = (I-Q)\left[g(x_1+x_2,\lambda)+\lambda x_2\right]$$

where Q is the projection from E onto the closed subspace $A(X_2)$, parallel to $\ker A^\alpha$. Applying K to the first equation, one has for λ small, $|\lambda| < \dfrac{1}{||KQ||}$.

$$(4)\quad \begin{cases} x_2 - (I-\lambda KQ)^{-1}KQg(x_1+x_2,\lambda) = 0 \\ \text{i.e.}\quad x_2 - F(x_1,x_2,\lambda) = 0 \end{cases}$$

Note that it will be shown in section V that one can take X_2 to be $D(A) \cap R(A^\alpha)$ so that $(I-Q)x_2 = 0$. From (1.6) and for $||x_1+x_2||_B$, $||x_1+x_2'||_B \leq r < r_o$, $|\lambda| < \min\left(\dfrac{1}{2||KQ||}, \lambda_o\right)$

then $||F(x_1,x_2,\lambda) - F(x_1,x_2',\lambda)|| < C_1M(r)||x_2-x_2'||_B$.

So for r so small that $C_1M(r) \leq 1/2$ (for x_1 fixed with $||x_1||_B < r/4$) $F(x_1,x_2,\lambda)$ will be a contraction from the open ball $\{x_2 \in X_2, ||x_2||_B < 3r/4\}$ into itself and yield a unique fixed point, provided the iteration is convergent. For this, it is enough to have:

$$||F(x_1,0,\lambda)||_E \leq C_1(||x_1||_B^2 + |\lambda|^{2\alpha+1}) \leq 3r/8,$$

which is easy to achieve by making r and λ small enough. Denote the solution of (4) by $x_2(x_1,\lambda)$ for $||x_1||_B < r_1$, $|\lambda| < \lambda_1$. Then $x_2(x_1,\lambda)$ is a continuous function of x_1 and λ. In fact:

$$||F(x_1,x_2,\lambda) - F(x_1',x_2',\lambda')|| \leq$$

$$||(I-\lambda KQ)^{-1}KQ||\ ||g(x_1+x_2,\lambda) - g(x_1'+x_2',\lambda')|| +$$

$$||((I-\lambda KQ)^{-1} - (I-\lambda' KQ)^{-1})KQ||\ ||g(x_1'+x_2',\lambda')||$$

$$\leq C_1(M(r)||x_1+x_2-x_1'-x_2'||_B) + C_1|\lambda-\lambda'|$$

$$+ C_2|\lambda-\lambda'|(||x_1'+x_2'||^2 + |\lambda'|^{2\alpha+1}).$$

With a possible restriction of the neighborhood, so that $M(r)$ and $||x_2'||^2$ are small enough, one gets:

$$||x_2'(x_1',\lambda') - x_2(x_1,\lambda)||_B \leq C(M(r)||x_1-x_1'||_B + |\lambda-\lambda'|)$$

and similarly:

$$||x_2(x_1,\lambda)||_B \leq C(||x_1||^2 + |\lambda|^{2\alpha+1})$$

b) A second reduction

So the study of (3) is reduced to the bifurcation equation:

$$Ax_1 - \lambda x_1 - (I-Q)[g(x_1+x_2(x_1,\lambda),\lambda)+\lambda x_2] = 0 \tag{5}$$

One may consider on $\ker A^\alpha$ the euclidean norm and write x_1 as $x_1 = (y_1,\ldots,y_m)$, choosing a basis so that A is in Jordan form. Then A has $d = n(A)$ Jordan blocks of size $k_i, i=1,\ldots,d$ with $\sum_1^d k_i = m$, $\max k_i = \alpha$, (each block corresponds to one element of ker A and A is nilpotent on ker A^{α})(1). On a typical $k \times k$ block, (5) has the form:

$$\begin{aligned} y_2 - \lambda y_1 &= g_1(x,\lambda) \\ y_3 - \lambda y_2 &= g_2(x,\lambda) \\ &\cdots\cdots \\ y_k - \lambda y_{k-1} &= g_{k-1}(x,\lambda) \\ -\lambda y_k &= g_k(x,\lambda) \end{aligned} \tag{5'}$$

where $g_1,\ldots,g_k$ are the components of $(I-Q)[g(x_1+x_2(x_1,\lambda),\lambda)+\lambda x_2]$ on the block. As above $y_2 = \lambda y_1 + g_1(x,\lambda)$ can be solved uniquely for y_2, in terms of y_1,λ and the other coordinates, as a function $y_2(y_1,\lambda,y_3,\ldots,y_k\ldots)$ continuous in its arguments, with the property that:

$$|y_2| \leq C(|\lambda y_1| + ||(y_1,y_3,\ldots,y_m)||^2 + |\lambda|^{2\alpha+1})$$

if $|y_\ell| < r$ for $\ell = 1,3,4,\ldots,m$ and $|\lambda| < \lambda_o$.

Similarly $y_3 = \lambda y_2 + g_2 = \lambda^2 y_1 + \lambda g_1(y_1,y_3,y_4,\ldots,y_m,x_2,\lambda) + g_2$ can be solved by contraction with $|y_3| \leq C(|\lambda^2 y_1| + ||(y_1,y_4,\ldots,y_m)||^2 + |\lambda|^{2\alpha+1})$. So the block reduces to:

$-\lambda^k y_1 = \lambda^{k-1} g_1 + \ldots + \lambda g_{k-1} + g_k = -\tilde{g}_1(y_1,\ldots,\lambda)$ where $\tilde{g}_1$ is function of y_1 and of the coordinates not appearing in this block, with the smallness and the continuity properties of g. Hence, after a similar reduction for all blocks and, if P is the canonical projection on ker A, $(I-P)x_1 = y$ and $Px_1 = z$, y is then defined as a continuous function of z and λ with

---

(1) E.N. Dancer has used a similar technique in [9] and [10].

$$||y|| < C(||z||^2 + |\lambda|^{2\alpha+1} + ||\lambda z||).$$

It follows that (3) is equivalent to:

$$(6)\quad \begin{cases} \lambda^{k_1} z_1 - \tilde{g}_1(z,y(z,\lambda),x_2(z,\lambda),\lambda) = 0 \\ \cdots\cdots\cdots\cdots\cdots\cdots\cdots\cdots \\ \lambda^{k_d} z_d - \tilde{g}_d(z,y(z,\lambda),x_2(z,\lambda),\lambda) = 0 \end{cases}$$

where $\tilde{g} = (\tilde{g}_1,\ldots,\tilde{g}_d)$ verifies:

$$||\tilde{g}(z,y,x_2,\lambda)|| \leq C(||z+y+x_2||_B^2 + |\lambda|^{2\alpha+1} + |\lambda|\ ||x_2||)$$

that is: $||\tilde{g}(z,\lambda)|| \leq C(||z||_B^2 + |\lambda|^{2\alpha+1})$

for some constant C, using the estimates for y and x_2 and for $||z||_B$ and $|\lambda|$ accordingly small.

Note that if $|\lambda| < 1$ the vector $(\lambda^{k_1} z_1,\ldots,\lambda^{k_d} z_d)$ has norm $||(\lambda^{k_1} z_1,\ldots,\lambda^{k_d} z_d)|| \geq |\lambda|^\alpha ||z||$ since $k_i \leq \alpha$.

Equation (6) will be studied in the set:

$$D = \{(z,\lambda) \in \ker A \times \mathbb{K} \ / \ ||z||^2 + |\lambda|^2 \leq r^2 + (Mr)^{2/\alpha}\}$$

with

$M = 2C + 1$ (C the above constant).

and

$$r \leq r_o \leq \min\ (\text{diameter of the neighborhoods used above},\ M^{-(2\alpha+1)})$$

On the boundary of D, (6) with the side condition:

$$||z||^2 = r^2 \quad (\text{i.e. } |\lambda|^2 = (Mr)^{2/\alpha})$$

has no solution, since:

$$|\lambda^\alpha|\ ||z|| = Mr^2 > C(r^2+(Mr)^2(Mr)^{1/\alpha}) \geq ||\tilde{g}(z,\lambda)||$$

which is true, by the choice of M and r, and moreover $|\lambda|^\alpha = Mr < M^{-2\alpha} < 1$ justifying the estimation $|\lambda|^{k_i} > |\lambda|^\alpha$.

Proof of i): Fix λ such that: $|\lambda| = (Mr)^{1/\alpha}$ then (6), on the sphere $||z||^2 = r^2$, can be deformed to $(\lambda^{k_1} z_1, \dots, \lambda^{k_d} z_d)$. It is easy to compute the degree of this mapping which is 1 in the complex case and sign λ^m in the real case. Theorem 2.2. shows then the existence of the trivial solutions.

Proof of ii) and iii): Adding to (6) the equation $||z||^2 - r^2$, one can then deform, on the boundary of D, the system to the equations (2.5). Theorem 2.6 gives the desired result.

Theorem 3.4: If $\dim_{\mathbb{K}} \ker A = 1$ and given the hypothesis of theorem 3.2, then for $\mathbb{K} = \mathbb{R}$ and m odd: There are at least two bifurcating solutions, namely for $Px = \pm r$.

For $\mathbb{K} = \mathbb{C}$ and any m, (3) has bifurcating solutions in all directions Px, with $||Px|| = r$.

Note: A sharper result can be obtained, when $m = 1$ and with the first inequality of (1.6) replaced by:

$$||g(x,\lambda) - g(x',\lambda')||_E \leq M(r)||x - x'||_B + C|\lambda - \lambda'|(|\lambda|^2 + |\lambda'|^2)$$

for $||x||_B, ||x'||_B \leq r$ and $|\lambda|, |\lambda'| \leq \lambda_o$.

Rabinowitz [27], Nirenberg [25].

In fact let x_o be in $\ker A$ such that: $||x_o|| = 1$. Then set $z = \mu x_o$ and (6) reduces to $\lambda - \tilde{g}(\mu,\lambda)/\mu = 0$. A contraction argument gives a unique solution $\lambda(\mu)$ continuous in μ with $\lambda(0) = 0$. So the non-trivial branch is of the form:

$$(x,\lambda) = (\mu x_o + x_2(\mu x_o, \lambda(\mu)), \lambda(\mu))$$

Note also (in the real case) that if $\tilde{g}$ is C^2 with: $\tilde{g}_\mu(0,0) = \tilde{g}_\lambda(0,0) = \tilde{g}_{\lambda\lambda}(0,0) = \tilde{g}_{\mu\lambda}(0,0) = 0$, then one can show (Nirenberg

[25]), using the Morse Lemma, that the set of solutions of (6) consists of two continuous curves, one corresponding to the trivial solutions the other to the bifurcating solutions. If $\tilde{g}$ is smoother, these curves will intersect transversally at $(0,0)$.

Proof of Theorem 3.4: (6) reduces here to $\lambda^m z - \tilde{g}(z,\lambda) = 0$. Fix $z = z_o$ such that $|z_o| = r$, and consider the equation (6) on the set: $\{\lambda \in K \,/\, |\lambda| \leq (Mr)^{1/\alpha}\}$. On the boundary of this set one can deform (6) to $\lambda^m z_o$, which, for the real case as a map from S^o into $\mathbb{R}-\{0\}$, has degree different from zero only if m is odd, and in the complex case as a map from S^1 into $\mathbb{R}^2-\{0\}$ has degree m.

Remark 3.5: It is interesting to note the relationship between the growth allowed for g and the different terms in the Jordan form. In fact, in $(5)'$ it is enough to assume that: $|g_i(x_1,\lambda)| < C(||x_1||^2 + |\lambda|^{\alpha+i+1})$. Then it is easy to see that in (6), $\lambda^{k_j} z_j - \tilde{g}_j(z,\lambda)$ has the estimate: $|\tilde{g}_j(z,\lambda)| < C(||z||^2 + |\lambda|^{\alpha+k_j+1})$. Replacing the usual norm on ker A by the equivalent norm $||z|| = \max|z_i|$, on the boundary of the ball $\{(z,\lambda)/||z||^2 + |\lambda|^2 < r^2 + (Mr)^{2/\alpha}\}$, (6) and the side condition $||z||^2 - r^2$ have no zero: In fact if $|z_i| = r$ for some i, then $|\lambda^{k_i} z_i| = (Mr)^{k_i/\alpha} r$, while $|\tilde{g}_i(z,\lambda)| \leq C(r^2 + (Mr)^{1+\frac{k_i}{\alpha}+\frac{1}{\alpha}})$. So, since $k_i \leq \alpha$, $(Mr)^{k_i/\alpha} r$ is the dominant term for suitable choices of M and r.

Remark 3.6: It is also worthy to note that the condition $||g(x,\lambda)|| < C(||x||^2 + |\lambda|^{2\alpha+1})$ is, in a sense, sharp. In fact the system:

$$y_2 - \lambda y_1 = 0$$

$$\cdots\cdots\cdots\cdots$$

$$y_k - \lambda y_{k-1} = 0$$

$$- \lambda y_k = |y_1|^2 + |\lambda|^{2k}$$

has no solution except $\lambda = 0$, $z = 0$, since it reduces to: $-\lambda^k y_1 = |y_1|^2 + |\lambda|^{2k}$, and taking absolute values on both sides yields the result.

<u>Remark 3.7</u>: If in (1.6) $M(r)$ is of the form Cr (which is the case if g is in C^2, $g(0,0) = g_x(0,0) = 0$), then the set of trivial solutions can be defined uniquely as a continuous function of λ, with $|\lambda| = (Mr)^{1/\alpha}$ for M large enough.

Since $z_1 = \lambda^{-k_1}\tilde{g}_1, z_2 = \lambda^{-k_2}\tilde{g}_2, \ldots, z_d = \lambda^{-k_d}\tilde{g}_d$, gives for a fixed λ: $||z-z'|| \leq |\lambda|^{-\alpha}||\tilde{g}(z,\lambda)-g(z',\lambda)|| \leq \frac{Cr}{Mr}||z-z'||$.

So for $M > 2C + 1$, a contraction yields a unique solution with:

$$||z(\lambda)-z(\lambda')|| \leq |\lambda|^{-\alpha}||\tilde{g}(z,\lambda)-\tilde{g}(z',\lambda)||$$

$$+ D|\lambda-\lambda'| \frac{(||z'||^2+|\lambda'|^{2\alpha+1})}{|\lambda|^{\alpha}\ |\lambda'|^{\alpha}}$$

$$\leq \tfrac{1}{2}||z(\lambda)-z(\lambda')|| + \tfrac{1}{2}C|\lambda-\lambda'|$$

hence the continuity of z with respect to λ follows.

The direct proof for theorem 3.2 permits one to give several immediate consequences, such as:

<u>Corollary 3.8</u>: <u>If in (6), one of the $\tilde{g}$'s is of the form</u> $\lambda\tilde{h}$ <u>for some</u> $\tilde{h}$ <u>with</u> $|\tilde{h}(z,\lambda)| \leq C(||z||^2 + |\lambda|^{2\alpha})$, <u>then the conclusions of theorems 3.2 and 3.4 hold without any restriction</u>

<u>on the multiplicity</u> m.

<u>Proof</u>: If $\tilde{g}_i = \lambda\tilde{h}$, then (6) can be reduced to a system where the total power in λ appearing in (2.8) is odd. (Use Remark 3.5).

<u>Corollary 3.9</u>: <u>Assume</u> $||g(x,\lambda)|| \leq C(||x||^2 + |\lambda|^{2\alpha+3})$ <u>if</u> m <u>is even and</u> $||g(x,\lambda)|| \leq C(||x||^2 + |\lambda|^{2\alpha+1})$ <u>if</u> m <u>is odd</u>.

<u>Suppose there exists</u> r, $r < r_o$, <u>and some integer</u> ℓ, $1 \leq \ell \leq d$, <u>such that</u>: In (6) $\tilde{g}_\ell\ (z,0) \neq 0$ <u>for all</u> z <u>with</u> $||z|| = r$.

<u>Then if</u> $\dim_K \ker A > 1$, <u>the conclusions of theorem 3.2 hold for all multiplicities and the non-trivial solutions occur with</u> λ <u>different from zero</u>.

<u>Proof</u>: Suppose $\ell = 1$ and replace (6) by:

$$\lambda^{k_1} z_1 - \tilde{g}_1$$

$$\lambda^{k_2+\varepsilon} z_2 - \lambda^{\varepsilon}\tilde{g}_2$$

$$\lambda^{k_3} z_3 - \tilde{g}_3 \quad \text{where } \varepsilon = 0 \text{ if } m \text{ is odd}$$

$$\cdots\cdots\cdot \qquad \varepsilon = 1 \text{ if } m \text{ is even}$$

$$\lambda^{k_d} z_d - \tilde{g}_d$$

$$||z||^2 - r^2$$

Then this system has a solution, by theorem 3.2, with $\lambda \neq 0$ since $\tilde{g}_1(z,0) \neq 0$, and hence gives a solution to (6).

<u>Remark 3.10</u>: If one of k_i's, for i different from ℓ, is such

that: $k_i < \alpha$, than g may satisfy the estimate of theorem 3.2.

Example 3.11: Suppose E is a Hilbert space, and that A is self-adjoint. Assume $|(g(z,0),v_\ell)| \geq k||z||^2$ for all z in $\ker A$ with $||z|| \leq r_o$ and some v_ℓ in $\ker A$. Then $\tilde{g}_\ell(z,0) = (g(z+x_2(z,0),0),v_\ell)$ with $x_2(z,0) = o(||z||)$ satisfies the hypothesis of corollary 3.9 for r small enough.

Remark 3.12: A combination of the two corollaries may be useful to establish the existence of solutions (x,λ) with $x \neq 0$, $\lambda \neq 0$ in the case where $g = \lambda h$, h satisfying the estimates.

Example 3.13: In the even multiplicity case it is possible to have an equation of bifurcation with no solution but the trivial solution: If λ, x_1, x_2 are real and k_1, k_2 are positive integers with $k_1 = k_2 + 2n$ for some non-negative n, then the equations:

$$\left.\begin{aligned} \lambda^{k_1} x_1 + t\, x_2 &= 0 \\ \lambda^{k_2} x_2 - t\, x_1 &= 0 \end{aligned}\right\} \quad \text{with} \quad t = x_1^2 + x_2^2$$

have the only solution $x_1 = x_2 = 0$.

Example 3.14: For the complex case, let $t = |z_1|^2 + |z_2|^2$ and consider:

$$\left\{\begin{aligned} &\lambda^{k_1} z_1 + t(\lambda^n \bar{z}_2 - nt\bar{z}_1 + n\,\frac{\lambda^{k_2+n}}{|\lambda|^{n+\varepsilon}} z_2) = 0 \\ &\lambda^{k_2} z_2 - t(\frac{\bar{\lambda}^n}{|\lambda|^{n-\varepsilon}} \bar{z}_1 + nt\bar{z}_2 + n\,\lambda^{k_2+n} z_1) = 0 \end{aligned}\right.$$

where: $\varepsilon = 0$ if $n = 0$ (λ^o means 1)

$$\varepsilon = 1 \quad \text{if} \quad n > 0 \quad \text{and} \quad k_2 \geq 2$$

$$\varepsilon = \frac{1}{2} \quad \text{if} \quad n > 0 \quad \text{and} \quad k_2 = 1.$$

In order to see that the only solution of this pair of equations is the trivial one, it is enough to note that they result from the application of the invertible (for $t \neq 0$) matrix:

$$\begin{bmatrix} \lambda^n & nt \\ -nt & \dfrac{\bar{\lambda}^n}{|\lambda|^{n-\varepsilon}} \end{bmatrix}$$

to the vector $(\lambda^{k_2+n} z_1 + t\bar{z}_2, \dfrac{\lambda^{k_2+n}}{|\lambda|^{n+\varepsilon}} z_2 - t\bar{z}_1)$.

As it was proved in Remark 2.9, this vector is zero only if $z_1 = z_2 = 0$.

Note that although the nonlinear part is continuous and satisfies the smallness condition of (1.6) it does not satisfy (when $k_2=1$) the Lipschitz condition in λ. It would be interesting to construct an example with smoother non-linear part. Note that it is easy to construct examples from the above for higher dimensional kernels.

§ IV : BIFURCATION FOR FREDHOLM OPERATORS OF NON NEGATIVE INDEX.

The method of reduction used in the preceding section, although useful in most applications because of the simple form of the bifurcation equation (6), is not suitable to study the more complicated equations which will be treated in section VI of this chapter and in the following chapter. The reduction to the finite system (6) is in fact a special case of the method which will be developed in this section.

Given B and E as in I, A is a Fredholm operator from B to E with non-negative index i(A). The aim of this section is to give sufficient conditions for the existence of bifurcation for equation

$$(2) : \quad Ax - \lambda x = g(x,\lambda),$$

where g satisfies the conditions of I.

a) <u>A new reduction</u>

Ker A being a finite dimensional subspace of B, there is a continuous projection P from B onto ker A and one can decompose B as:

$$B = \ker A \oplus B_2,$$

where B_2 is a closed subspace of B.
Then: $x = x_1 + x_2$ with x_1 in ker A and x_2 in B_2.
Similarly, R(A) being closed and finite codimensional, let Q be a projection from E onto R(A). Then ker Q $\equiv$ coker A is a complementing subspace of dimension $d(A) = d^*$. Since A restricted to B_2 is a one to one, onto and bounded operator, the open mapping theorem ascertains the existence of a continuous operator K from R(A) onto B_2 such that: $AKQ = Q$ and $KA(I-P) = I - P$; moreover K depends uniquely on P and Q.

Then (2) is equivalent to:

$$\begin{cases} Ax_2 - \lambda Q(x_1+x_2) - Qg(x_1+x_2,\lambda) = 0 \\ \lambda(I-Q)(x_1+x_2) + (I-Q)g(x_1+x_2,\lambda) = 0 \end{cases}$$

Applying K to the first equation yields:

$$x_2 - \lambda KQx_2 - \lambda KQx_1 - KQg(x_1+x_2,\lambda) = 0$$

and for $|\lambda| < 1/\ ||KQ||$:

$$(7) \quad x_2 - (I-\lambda KQ)^{-1}\ \lambda KQx_1 - (I-\lambda KQ)^{-1}\ KQg(x_1+x_2,\lambda) = 0$$

For $||x||_B$ and $|\lambda|$ sufficiently small equation (7), as before equation (4), admits a unique solution $x_2(x_1,\lambda)$ with the properties:

$$\begin{cases} ||x_2(x_1,\lambda)-x_2(x_1',\lambda')||_B \leq \tilde{C}(M(r)||x_1-x_1'||_B+|\lambda-\lambda'|+|\lambda|\ ||x_1-x_1'||_B) \\ ||x_2(x_1,\lambda)||_B \leq \tilde{C}(||x_1||_B^2 + |\lambda|^{2n+1}+|\lambda|\ ||x_1||_B) \end{cases}$$

for $||x_1||_B$, $||x_1'||_B < r \leq r_o$ $|\lambda|,|\lambda'| < \lambda_o$.

And if $||x_1|| \leq r$, $|\lambda| \leq (Mr)^{1/n}$ for some M, then $||x_2(x_1,\lambda)||_B = o(r)$.

The insertion of $x_2(x_1,\lambda)$ in the second equation yields:

$$\lambda(I-Q)(x_1+(I-\lambda KQ)^{-1}\lambda KQx_1) + (I-Q)[g(x_1+x_2,\lambda) + (I-\lambda KQ)^{-1}\lambda KQg(x,\lambda)] = 0$$

Making use of the fact $I + (I-\lambda KQ)^{-1}\lambda KQ = (I-\lambda KQ)^{-1}$, one has the bifurcation equation:

$$(8) \quad \lambda(I-Q)(I-\lambda KQ)^{-1}x_1 + (I-Q)(I-\lambda KQ)^{-1}g(x_1+x_2(x_1,\lambda),\lambda) = 0$$

Given Euclidean norms on $\ker A$ and $\operatorname{coker} A$, with their respective bases, and denoting $(I-Q)(I-\lambda KQ)^{-1}\big|_{\ker A}$ by $A(\lambda)$, then $A(\lambda)$ is a $d^* \times d$ matrix whose elements are analytic functions of λ.

<u>Remark 4.1</u>: If one follows the calculation which ended in the system (6), one will recognize the above reduction with special

choices for P and Q. But in general the entries of $A(\lambda)$ are not polynomials in λ as the following example shows:

In $\mathbb{R}^2$ let $A = \begin{bmatrix} 0 & 1 \\ 0 & 0 \end{bmatrix}$. ker A is generated by $\begin{bmatrix} 1 \\ 0 \end{bmatrix}$ and coker A by $\begin{bmatrix} 0 \\ 1 \end{bmatrix}$. One can choose $P = \begin{bmatrix} 1 & -1 \\ 0 & 0 \end{bmatrix}$. Then $P\begin{bmatrix} x_1 \\ x_2 \end{bmatrix} = \begin{bmatrix} x_1 - x_2 \\ 0 \end{bmatrix}$ is a projection on ker A. Let $Q = \begin{bmatrix} 1 & 0 \\ 0 & 0 \end{bmatrix}$ then $K = \begin{bmatrix} 1 & 0 \\ 1 & 0 \end{bmatrix}$ and $(KQ)^n = \begin{bmatrix} 1 & 0 \\ 1 & 0 \end{bmatrix}$ for all n, $n \geq 1$. Then an easy computation gives: $\lambda A(\lambda)x_1 = \frac{\lambda^2}{1-\lambda} x_1$. While if P is chosen to be : $\begin{bmatrix} 1 & 0 \\ 0 & 0 \end{bmatrix}$ then $K = \begin{bmatrix} 0 & 0 \\ 1 & 0 \end{bmatrix}$ with the same Q. Then $(KQ)^2 = 0$ and $\lambda A(\lambda)x_1 = \lambda^2 x_1$.

b) <u>Isolated eigenvalue</u>.

<u>Remark 4.2</u>: It is a well known fact (Goldberg [14], p.98-116), that if A is a Fredholm operator, than its minimum modulus $\gamma(A)$ is positive and that for $|\lambda| < \gamma(A)$, $A - \lambda I$ is also a Fredholm operator of same index $i(A)$. So if $i(A)$ is non-negative the spectrum of A, near 0, will be formed purely of eigenvalues, since either $n(A-\lambda I) = 0$ and so $i(A) = 0$ and λ is in the resolvent set of A, or $n(A-\lambda I) > 0$ and λ is an eigenvalue. This shows that the local study of the spectrum of A is entirely equivalent to knowing the dimension of ker $A(\lambda)$ for $\lambda \neq 0$, since if g is zero (7) and (8) become:

$$\begin{cases} x_2 - (I-\lambda KQ)^{-1} \lambda KQ\, x_1 = 0 \\ \lambda (I-Q)(I-\lambda KQ)^{-1} x_1 = 0 \end{cases}$$

These statements will be made more precise and proved in section VI for the more general case of $A - T(\lambda)$.

Hence a natural assumption is that 0 is, in some sense, isolated in the spectrum of A: more precisely, assume that there exists $\lambda_o > 0$ such that for all λ, with $0 < |\lambda| < \lambda_o$, then:

$$\text{codim Range}(A-\lambda I) = d(A-\lambda I) = 0 \tag{9}$$

By remark 4.2, this assumption is equivalent to $A(\lambda)$ having maximal rank.

Remark 4.3: In III, for $g = 0$, equations (4) and (5)' reduce to $x_2 = 0$, $y_2 = \lambda y_1, \ldots, y_k = \lambda y_{k-1}, \lambda y_k = 0$. This implies that 0 is an isolated eigenvalue of A.

c) Study of the bifurcation equation (8)

Now if assumption (9) holds, it follows that for each λ, $\lambda A(\lambda)$ has a $d^* \times d^*$ minor whose determinant is different from zero at λ. Since the determinants of the $d^* \times d^*$ minors are analytic functions of λ, vanishing at 0, they cannot accumulate zeros near 0, except if they vanish identically. Hence, because one of them is not zero for each λ, one can assume there is a fixed $d^* \times d^*$ minor of $\lambda A(\lambda)$ with determinant $\Delta(\lambda) = \lambda^m a(\lambda)$ with $a(\lambda)$ analytic in λ and $a(0) \neq 0$.
This m will be called the algebraic multiplicity of the eigenvalue 0.

Suppose that on $\ker A$, the above minor corresponds to the first d^* coordinates and write $x_1 = (y_1, \ldots, y_{d^*}, z_1, \ldots, z_i)$ with $i = i(A)$.

Theorem 4.4: Assume that $g(x,\lambda)$ satisfies condition (1.6) with

$n = m - d^* + 1$, _where_ m _is the algebraic multiplicity_. _Then_:

i) _If_ $i(A) = 0$, _the conclusions of theorems 3.2 and 3.4 hold_.

ii) _If_ $i(A) > 0$, _there is always bifurcation_. _In fact for any arbitrary constants_ $\varepsilon_1, \ldots, \varepsilon_i$, _with_ $\sum_1^i |\varepsilon_j|^2 = o(r^2)$ r _small enough_, $Ax - \lambda x = g(x,\lambda)$ _has a solution for_ $|\lambda| < (Mr)^{1/n}$, $||x_1|| = r$ _and_ $\lambda^k z_1 = \varepsilon_1$, $z_2 = \varepsilon_2, \ldots, z_i = \varepsilon_i$ _where_ $k = 0$ _if_ m _is odd_, $k = 1$ _if_ m _is even_.

Remark 4.5: (ii) is not surprising since this is what one should expect from the linear part of (8) if g were a constant.

Proof of Theorem 4.4: The goal of the proof is to reduce (8) to the system (2.5) in order to apply theorem 2.6.

As the statement of the theorem shows, the $d^* \times d$ system in (8) is first replaced by a $d \times d$ system by adding the equations:

$$\lambda^k z_1 - \varepsilon_1 = 0,\ z_2 - \varepsilon_2 = 0, \ldots, z_i - \varepsilon_i = 0$$

So one expands (8) to:

$$\left[\begin{array}{c|ccccc} \lambda\Delta_1(\lambda) & \lambda\Delta_2(\lambda) & & & & \\ \hline & \lambda^k & & & & 0 \\ & & 1 & & & \\ 0 & & & 1 & & \\ & & & & 1 & \\ & & & & & 1 \end{array}\right] \left[\begin{array}{c} y_1 \\ \vdots \\ y_{d^*} \\ \hline z_1 \\ \vdots \\ z_i \end{array}\right] + \left[\begin{array}{c} (I-Q)(I-\lambda KQ)^{-1} g \\ \hline \varepsilon_1 \\ \vdots \\ \varepsilon_i \end{array}\right]$$

where $\Delta_1(\lambda)$ is the $d^* \times d^*$ minor with determinant $\lambda^m a(\lambda)$. This new matrix has determinant $\lambda^{m+k} a(\lambda)$, where $m + k$ is odd, which will insure the existence of a non-trivial solution provided this new system is reduced to (2.5). So, only the case $i(A) = 0$ will be considered.

Since $\lambda A(\lambda)$ is non-singular for $0 < |\lambda| < \lambda_o$ one has:

$$||\lambda A(\lambda)x_1|| \geq \frac{||x_1||}{||(\lambda A(\lambda))^{-1}||}$$

Now $||(\lambda A(\lambda))^{-1}|| \leq C \max_{i,j}|b_{ij}(\lambda)|$ where C depends only on d^* and $b_{ij}(\lambda)$ are the entries of $(\lambda A(\lambda))^{-1}$. Since each term of $\lambda A(\lambda)$ begins with λ, for λ small enough:

$$|b_{ij}(\lambda)| \leq K \frac{|\lambda|^{d^*-1}}{|\lambda|^m}$$

for some constant K.

So $||\lambda(A(\lambda)x_1|| \geq ||x_1|| \frac{|\lambda|^{m-d^*+1}}{KC} = \tilde{C}|\lambda|^n \; ||x_1||$.

(It is easy to check that this estimate holds for the positive index case). But the right hand side of (8) has a norm:

$$||(I-Q)(I-\lambda KQ)^{-1}g(x_1+x_2(x_1,\lambda),\lambda)|| \leq C(||x_1+x_2||_B^2 + |\lambda|^{2n+1})$$

So on the sphere $||x_1||^2 + |\lambda|^2 = r^2 + (Mr)^{2/n}$, M large enough and r small, with this estimate together with the estimate for $||x_2||_B$, the argument used in the proof of theorem 3.2 enables one to deform (8), with the additional condition $||x_1||^2 - r^2$, to its linear part. Note that, by the choice of $\varepsilon_j, j = 1,\ldots,i$, this deformation is possible in the positive index case.

So, in order to complete the proof, one has to study the homotopy type of $\{\lambda A(\lambda)x_1, \; ||x_1||^2 - r^2\}$ on the sphere $||x_1||^2 + |\lambda|^2 = r^2 + (Mr)^{2/n}$.

Develop the determinant of $\lambda A(\lambda)$ with respect to the first row: $\lambda^m a(\lambda) = \sum_{i=1}^{j} \lambda^{k_i} a_i(\lambda)$ with $a_i(0) \neq 0$, $a(0) \neq 0$, $1 \leq j \leq d^*$, $a_i(\lambda)$ analytic in λ and $\lambda^{k_i} a_i(\lambda)$ corresponds to $\lambda a_{1i}(\lambda)\Delta_{1i}(\lambda)$, $\Delta_{1i}(\lambda)$ the attached cofactor.

Let $k_o = \min_{i,1,\ldots,j} k_i \leq m$

and let $i_1,\ldots,i_\ell$ correspond to this exponent. i.e.:

$$\lambda^m a(\lambda) = \lambda^{k_o}\left(\sum_{s=1}^{\ell} a_{i_s}(\lambda)\right) + \sum_{i \neq i_1,\ldots,i_\ell} \lambda^{k_i} a_i(\lambda)$$

Now $a_{i_s}(0) = r_{i_s} e^{i\theta_{i_s}}$ with $0 \leq \theta_{i_s} < 2\pi$

and $a(0) = r_o e^{i\theta_o}$ $0 \leq \theta_o < 2\pi$.

<u>Claim</u>: One of the i_s' is such that $\theta_{i_s} \neq \theta_o + \pi$.

If not $a_{i_s}(0) = -r_{i_s} e^{i\theta_o}$. So if $k_o < m$, one needs $\sum_{s=1}^{\ell} a_{i_s}(0) = 0$

which is not possible, and if $k_o = m$, one needs $\sum_{s=1}^{\ell} a_{i_s}(0) = a(0)$

and the claim is proved.

So let i_o be such that $\theta_{i_o} \neq \theta_o + \pi$ (For $\mathbb{K} = \mathbb{R}, \theta_{i_o} = \theta_o$). Replace then the elements $a_{1j}(\lambda)$ of the first row by:

$(1-t)a_{1j}(\lambda)$ for $j \neq i_o$ and by $((1-t)+t\lambda^{m-k_o})a_{1i_o}(\lambda)$ for $j=i_o$.

The new matrix $\lambda A(\lambda,t)$ will have a determinant:

$$(1-t)\lambda^m a(\lambda) + t\lambda^m a_{i_o}(\lambda) = \lambda^m((1-t)a(0) + ta_{i_o}(0) + \lambda(\ldots))$$

By choice of i_o : $(1-t)a(0) + ta_{i_o}(0) \neq 0$ for all t in $[0,1]$, and if one takes λ small enough this term will be dominant, so that the determinant will not vanish except at 0. (Note that these estimations for λ do not depend on the nonlinearity g and so can be performed before deforming (8) to a simpler system). The deformation is allowable since one needs that $\{\lambda A(\lambda,t)x_1,\ ||x_1||^2 - r^2\}$ be not zero on the sphere, and the first equation vanishes only at $\lambda = 0$, the second only for $|\lambda|=(Mr)^{1/n}$. So one has replaced $\lambda A(\lambda)$ by a matrix which has only one non-zero element in the first row, $\lambda^{m-k_o} a_{1i_o}(\lambda)$. Moreover replacing $a_{ji_o}(\lambda)$, for $j \neq 1$, by $(1-t)a_{ji_o}(\lambda)$ will not change the determinant, so the jth column disappears and $\det(\lambda A(\lambda)) = \lambda^{m-k_o} a_{1i_o}(\lambda)\Delta_1(\lambda)$. A repetition of the above arguments permits us to reduce the matrix $\lambda A(\lambda)$ to the matrix $B(\lambda)$ which has exactly one element in each row and each column, with $\det(B(\lambda))= \lambda^m b(\lambda)$. (For $\mathbb{K} = \mathbb{R}$, $b(0)$ has the sign of $a(0)$):

$$B(\lambda)x_1 = (a_1(\lambda)y_{i_1},\ldots,a_d(\lambda)y_{i_d}) \quad \text{with} \quad a_j(\lambda)=\lambda^{n_j}b_j(\lambda), b_j(0)=b_j\neq 0$$

$$\text{and} \quad \sum_1^d n_j = m.$$

A final deformation, again choosing λ small enough, will keep only the lowest order term in $a_j(\lambda)$, i.e. $\lambda^{n_j}b_j$. At last, a series of rotations of the form:

$$(t\lambda^{n_j}y_{i_j}b_j + (1-t)b_k\lambda^{n_k}y_{i_k},\ -(1-t)\lambda^{n_j}y_{i_j}b_j + tb_k\lambda^{n_k}y_{i_k})$$

will hold:

$$(B(\lambda)x_1,||x_1||^2-r^2) \sim (b_1\lambda^{k_1}y_1,\ldots,b_d\lambda^{k_d}y_d,||x_1||^2-r^2)$$

with $\sum_1^d k_i = m$.

For $\mathbb{K} = \mathbb{R}$, this map has degree: $-2 \operatorname{sign} (\prod_1^d b_i)$ if m is odd and 0 if m is even. For $\mathbb{K} = \mathbb{C}$ the class of this map is $\Sigma(-m\eta)$. This completes the proof of theorem 4.4.

An immediate question is how the algebraic multiplicity depends on the choices of the projections. An answer to this is:

Theorem 4.5: *Let* $i(A) = 0$, 0 *is an isolated eigenvalue of* A *and let* $\det(\lambda A(\lambda)) = \lambda^m a(\lambda)$. *Then*:

i) *If* $\mathbb{K} = \mathbb{R}$: *the parity of* m *and the sign of* $a(0)$ *are independent of* P *and* Q.

ii) *If* $\mathbb{K} = \mathbb{C}$: m *is independent of* P *and* Q.

Proof: Let P_o, P_1 be two projections of B onto $\ker A$ and Q_o, Q_1 be two projections of E onto $R(A)$.

Let: $P_t = tP_1 + (1-t)P_o$, $Q_t = tQ_1 + (1-t)Q_o$.

Using the fact that $P_1P_o=P_o, P_oP_1=P_1$ and similar relations for Q_o and Q_1, it is easy to check that P_t (resp. Q_t) are continuous projections from B (resp. E) onto $\ker A$ (resp. $R(A)$) and that if one defines $X_t = \ker P_t, Y_t = \ker Q_t$ then: $B = \ker A \oplus X_t$, $E = R(A) \oplus Y_t$. So if $K_t = tK_1 + (1-t)K_o$, for each fixed t, K_t has the properties: $AK_tQ_t = Q_t, K_tA = I - P_t$. Then:

$$\det(\lambda(I-Q_t)(I-\lambda K_tQ_t)^{-1}\Big|_{\ker A}) = \det(\lambda A(\lambda,t)) = \sum_1^\infty \lambda^i a_i(t) = \lambda^{i_o} a_{i_o}(t) + \dots$$

where the series converges for small λ and, by construction, $a_i(t)$ are polynomials in t. Let

$$C(\lambda,t) = \lambda^{1-i_o}\det(\lambda A(\lambda,t)) = a_{i_o}(t)\lambda + \dots$$

It has been seen, for a fixed t, that 0 is an isolated eigenvalue if and only if $C(\lambda,t)$ is not identically zero. But for t_o in $[0,1]$ and for $\varepsilon > 0$ given, $|C(\lambda,t_o)| > 0$ for all λ with $\varepsilon \leq |\lambda| \leq \lambda_o = \frac{1}{2}$ radius of convergence of the series, since within the radius of convergence a zero of $C(\lambda,t_o)$ corresponds to an eigenvalue of A. By continuity, it is clear that the last statement is still true in some neighborhood of t_o. A covering argument will give $|C(\lambda,t)| > 0$ for all λ with: $\varepsilon \leq |\lambda| \leq \lambda_o$, for all t in $[0,1]$ and all $\varepsilon > 0$.

So $C(\lambda,t)$ defines a continuous map from $\{\lambda / |\lambda| \leq \lambda_o\} \times [0,1]$ into $\mathbb{K}$ vanishing only at $\lambda = 0$. So its index is well defined and by homotopy invariance of the index it is a constant which, for each t, is the index of the map defined by the leading term in the expression for $C(\lambda,t)$. For $\mathbb{K} = \mathbb{R}$, this index is: sign $a_{i_o}(t)$ if t is not a zero of $a_{i_o}(t)$; and if t_o is a zero of $a_{i_o}(t)$, so that $C(\lambda,t_o) = a_m(t_o)\lambda^{m-i_o+1} + \ldots$, the index is: sign $a_m(t_o)$ if $m - i_o + 1$ is odd or zero if $m - i_o + 1$ is even. This implies that m and i_o are of the same parity and that sign $a_{i_o}(t)$ is invariant.

For $\mathbb{K} = \mathbb{C}$ the index is one if t is not a zero of $a_{i_o}(t)$ and $m - i_o + 1$ otherwise. So one needs $m = i_o$.

<u>Remark 4.6</u>: Theorem 4.5 is valid only for the type of reduction used in (7) and (8). But it is clear that a change in the format of equations (7) and (8), as will be the case in chapter III, may change the sign of the degree but not its absolute value.

§ V : ISOLATED EIGENVALUE VERSUS FINITE MULTIPLICITY.

It has been seen in Remark 4.3 that if one assumes hypotheses 1) and 2) at the beginning of section III (where the natural algebraic multiplicity was used), then 0 is an isolated eigenvalue of A, i.e. hypothesis (9) is satisfied. It is also well known (Gokhberg-Krein [15] Theorem 4.2, Taylor [37] Theorem 9.6) that if 0 is an eigenvalue with finite multiplicity of a Fredholm operator on a complex Banach space E, then E admits the decomposition $\ker A^{\alpha} \oplus R(A^{\alpha})$. However, since this fact relies on a spectral decomposition of E given by the Cauchy formula (Kato [20]) a similar analysis cannot be carried out in the real case, unless one tries to complexify the spaces as in Dancer [9].

Note that if A is defined on E as a bounded operator of index 0, one may characterize finite multiplicity by showing that there is a bounded invertible operator B such that $BA^{\alpha} = A^{\alpha}B = I-K$ where K is compact: Schechter [31] Theorem 5.2.

The aim of this section is to prove the equivalence of these three concepts: finite multiplicity, isolated eigenvalue and decomposition of the space. So doing, a partial answer to a question asked by Taylor ([37] comment on Theorem 5.5) will be given.

Let A be a Fredholm operator of index zero with domain D(A) in E. Following Taylor [37], one defines:

$$D(A^n) = \{x \in E \ / \ x \in D(A),\ Ax \in D(A), \dots, A^{n-1}x \in D(A)\},$$

<u>Nullity of A</u>: $n(A) = \dim_{\mathbb{K}} \ker A$,

<u>Deficiency of A</u>: $d(A) = \operatorname{codim}_{\mathbb{K}} R(A)$,

Ascent of A: $\alpha(A) = \min(p \ / \ \ker A^p = \ker A^{p+1})$,

Descent of A: $\delta(A) = \min(p \ / \ R(A^p) = R(A^{p+1}))$.

Note that $\alpha(A)$ and $\delta(A)$ may be infinite.

Let Q be a projection of E onto $R(A)$ and, if $D(A)$ is given the graph norm $\| \ \|_B$ as in I, let P be the projection of $D(A)$ onto $\ker A$. Finally, let $K : E \to D(A) = B$ be such that $AKQ = Q$, $KA = I - P$.

Theorem 5.1: In this setting, the following assertions are equivalent:

i) for some finite α : $E = R(A^\alpha) \oplus \ker A^\alpha$.

ii) $\alpha(A) = \delta(A) < \infty$: finite multiplicity in the sense of III.

iii) 0 is an isolated eigenvalue of A.

Proof:

1) (i) $\to$ (ii)

$\ker A^\alpha \subset \ker A^{\alpha+\ell}$ and if $A^{\alpha+\ell} x = 0$ then for $\ell \leq \alpha$: $A^\alpha x \in \ker A^\ell \cap R(A^\alpha) \subset \ker A^\alpha \cap R(A^\alpha) = \{0\}$. And for $\ell \geq \alpha$: $A^\ell x \in \ker A^\alpha \cap R(A^\ell) \subset \ker A^\alpha \cap R(A^\alpha) = \{0\}$ so one can repeat this argument with ℓ replaced by $\ell - \alpha$. So $\ker A^\alpha = \ker A^{\alpha+\ell}$, hence $\alpha(A) \leq \alpha < \infty$. Then Taylor [37] Theorem 4.5 (c) proves the implication.

Note that $B = D(A) = \ker A^\alpha \oplus R(A^\alpha) \cap B$ where B has the graph norm.

2) (ii) $\to$ (iii)

As was noted in Remark 4.3, the decomposition of lemma 3.1 gives the system, (in the notation of section III which is different from the notation of the present section):

$$\begin{cases} x_2 - (I-\lambda KQ)^{-1}KQg(x) = 0 \\ Ax_1 - \lambda x_1 - \lambda(I-Q)x_2 - (I-Q)g(x) = 0. \end{cases}$$

So that if g is identically zero, this reduces to:

$$x_2 = 0,\ Ax_1 - \lambda x_1 = 0 \quad \text{on} \quad \ker A^{\alpha}.$$

But since A is nilpotent on $\ker A^{\alpha}$, the only solution to the last equation is $x_1 = 0$, i.e. $Ax - \lambda x = 0$ has no solution but 0, provided λ is small and not zero.

3) (iii) $\to$ (ii)

Recall that $Ax - \lambda x = 0$ is equivalent to:

$$(7) \quad x_2 - (I-\lambda KQ)^{-1}\ \lambda KQx_1 = 0$$

$$(8) \quad \lambda(I-Q)(I-\lambda KQ)^{-1}\ x_1 = \lambda A(\lambda)x_1 = 0$$

where $B = \ker A \oplus B_2$, P is the projection on $\ker A$, Q the projection of E onto $R(A)$ and K is the pseudo inverse of A restricted to B_2.

Lemma 5.2: Let $x \in E$ be such that:

$$(I-Q)(KQ)^{\ell}x = 0 \qquad \ell = 0,\dots,n-1$$

Let $y = (KQ)^n x$. Then $y \in D(A^n)$, $x = A^n y$, $P\,A^{\ell}y = 0$, $\ell = 0,\dots,n-1$ and conversely.

Proof: If $(I-Q)(KQ)^{\ell}x=0$ $\ell=0,\dots,n-1$ and $y = (KQ)^n x$ then $y \in B$, $Py = 0$.

Apply A: $Ay = AKQ(KQ)^{n-1}x = Q(KQ)^{n-1}x = (KQ)^{n-1}x$ since $(I-Q)(KQ)^{n-1}x = 0$. So $Ay \in B$ and $PAy = 0$. Applying A repeatedly one gets the first part of the statement.

Conversely: If $x = A^n y$ and $P A^{\ell} y = 0$ $\ell = 0,\ldots,n-1$ then $x \in R(A)$ i.e. $(I-Q)x = 0$.

Apply K : $KA(A^{n-1}y) = KQx = (I-P)A^{n-1}y = A^{n-1}y$. So $KQx \in R(A)$, and applying K repeatedly, one gets the conclusion.

<u>Lemma 5.3</u>: If $\alpha(A) = \infty$, then $\det \lambda A(\lambda) \equiv 0$.

<u>Proof</u>: Consider the sequence of finite dimensional subspaces:

$$\ker A \cap R(A^k) \subset \ker A \cap R(A^{k-1}) \subset \ldots \subset \ker A.$$

So there exists k_o such that $\ker A \cap R(A^k) = \ker A \cap R(A^{k_o})$ for all k, $k \geq k_o$. It is clear that if $\ker A \cap R(A^{k_o}) = \{0\}$ then $\alpha(A) \leq k_o$. ([37] Lemma 3.4). Hence if n_o is the dimension of $\ker A \cap R(A^{k_o})$ then n_o is positive.

Now, for any ℓ, $\ell \geq 0$, A^k maps $\ker A^{k+1} \cap R(A^{\ell})$ onto $\ker A \cap R(A^{k+\ell})$ with null-space $\ker A^k \cap R(A^{\ell})$. So for $k + \ell \geq k_o$:

$$(5.4) \qquad \dim \left[(\ker A^{k+1} \ominus \ker A^k) \cap R(A^{\ell})\right] = n_o$$

and since $R(A^{\ell})$ is contained in E and the following spaces are finite dimensional, one has:

$$(5.5) \quad (\ker A^{k+1} \ominus \ker A^k) \cap R(A^{\ell}) = \ker A^{k+1} \ominus \ker A^k$$

for any splitting of $\ker A^{k+1}$, independently of $R(A^{\ell})$. So according to (5.4) and (5.5), choose $y_1^{k_o+1},\ldots,y_{n_o}^{k_o+1}$ in $R(A) \cap \ker A^{k_o+1}$ spanning $\ker A^{k_o+1} \ominus \ker A^{k_o}$. From these elements, construct a Jordan form for A on $\ker A^{k_o+1}$ such that the associated projection P gives:

$$PA^{\ell} y_i^{k_o+1} = 0 \text{ for } \ell = 0,\ldots,k_o-1$$

$$PA^{k_o} y_i^{k_o+1} = A^{k_o} y_i^{k_o+1} \quad \text{for } \ell = k_o.$$

Recall that if F_{k_o+1} is the linear span of $y_1^{k_o+1}, \ldots, y_{n_o}^{k_o+1}$,

then: $A(F_{k_o+1}) \cap \ker A^{k_o-1} = \{0\}$ and A is one to one on

F_{k_o+1}. So if $F_{k_o} \equiv A(F_{k_o+1}) \oplus \langle y_{n_o+1}^{k_o}, \ldots, y_{n_{k_o}}^{k_o} \rangle$ is a comp-

lement to $\ker A^{k_o-1}$ in $\ker A^{k_o}$ and by induction, if

$F_{k_o+1-k} \equiv A(F_{k_o+1-k+1}) \oplus \langle y_{n_{k_o+2-k}+1}^{k_o+1-k}, \ldots, y_{n_{k_o-k+1}}^{k_o+1-k} \rangle$ is a comp-

lement to $\ker A^{k_o-k}$ in $\ker A^{k_o+1-k}$ and for $F_1 \equiv \ker A =$

$A(F_2) \oplus \langle y_{n_2+1}^1, \ldots, y_{n_1}^1 \rangle$ one has then, by construction:

$n_o \equiv n_{k_o+1} \leq n_{k_o} \leq \ldots \leq n_1 = n(A)$ and a basis for $\ker A^{k_o+1}$

is given by:

$$\left\{\begin{array}{lll}
y_1^{k_o+1}, \ldots, \; y_{n_o}^{k_o+1} & & \text{Basis for } \ker A^{k_o+1} \ominus \ker A^{k_o} \\
Ay_1^{k_o+1}, \ldots, \; Ay_{n_o}^{k_o+1}, & y_{n_o+1}^{k_o}, \ldots, y_{n_{k_o}}^{k_o} & \text{Basis for } \ker A^{k_o} \ominus \ker A^{k_o-1} \\
\vdots \qquad \vdots & & \\
A^{k_o} y_1^{k_o+1}, \ldots, A^{k_o} y_{n_o}^{k_o+1}, A^{k_o-1} y_{n_o+1}^{k_o}, \ldots & & \text{Basis for } \ker A.
\end{array}\right.$$

So if P_{k_o+1} is any projection from B onto $\ker A^{k_o+1}$, define

P on $\ker A^{k_o+1}$ as
$$\left.\begin{array}{ll} P A^{\ell} y_i^{k_o+1-k} = 0 & \text{if } \ell = 0, \ldots, k_o-k-1 \\ \qquad\qquad\; = A^{k_o-k} y_i^{k_o+1-k} & \text{for } \ell = k_o-k \end{array}\right\}$$

$k = 0,\dots,k_o$ and $i = n_{k_o-k+2} + 1,\dots n_{k_o-k+1}$ $(n_{k_o+2} = 0)$.

Then from the fact that $y_i^{k_o+1}$ are in $R(A)$, let $y_i^{k_o+2}$ be such that $Py_i^{k_o+2} = 0$ and $Ay_i^{k_o+2} = y_i^{k_o+1}$. So $y_i^{k_o+2}$ are in $\ker A^{k_o+2}$, are linearly independent and from (5.4), (5.5) form a splitting of $\ker A^{k_o+2} \ominus \ker A^{k_o+1}$. Moreover from (5.5) $y_i^{k_o+2}$ are in $R(A)$.

So one can give a basis for $\ker A^{k_o+3} \ominus \ker A^{k_o+2}$ by letting $y_i^{k_o+3}$ be such that $Py_i^{k_o+3} = 0$ and $Ay_i^{k_o+3} = y_i^{k_o+2}$, and so on...

Then let $x_i = A^{k_o} y_i^{k_o+1} = A^{k_o+1} y_i^{k_o+2} = \dots = A^{k_o+m} y_i^{k_o+m+1}$ for some $y_i^{k_o+m+1}$ with $PA^{\ell} y_i^{k_o+m+1} = 0$ for $\ell = 0,\dots,k+m-1$.

Hence by lemma 5.2: $(I-Q)(KQ)^{\ell} x_i = 0$ $\ell = 0,\dots,k+m-1$ and any m. So $\lambda A(\lambda)x_i \equiv 0$, which proves that 0 is not an isolated eigenvalue since $(x_i, x_2 = (I-\lambda KQ)^{-1} \lambda KQx_i)$ belongs to $\ker(A-\lambda I)$ for all λ small enough and i ranging from 1 to n_o, $n_o \geq 1$. Q.E.D. □□

The proof of theorem 5.1 will be completed when the implication iii) → i) is established, that is if 0 is an isolated eigenvalue then E has the splitting of (i). It is enough to show that A^{α}, α given by (ii), has index zero: In fact if A is a Fredholm operator, then A^{α} is a closed operator from its domain $D(A^{\alpha})$ into E (Goldberg [14]. Cor. IV.2.12) and since it is clear that $\ker A^{\alpha}$ (resp. $R(A^{\alpha})$) is finite dimensional (resp. finite codimensional), it follows that A^{α} is a Fredholm operator. Moreover one has: $R(A^{\alpha}) \cap \ker A^{\alpha} = \{0\}$ and $\alpha(A) = \delta(A)$, Taylor [37] theorems 3.7 (b) and 4.5 (c). So if one proves that

$i(A^{2^n}) = 0$ for all n then, for n so large that $\ker A^{2^n} = \ker A^\alpha$ and $R(A^{2^n}) = R(A^\alpha)$, one concludes that $i(A^{2^n}) = i(A^\alpha) = 0$. That is $\ker A^\alpha$ is a subspace of the right dimension to complement the closed subspace $R(A^\alpha)$ of E. This last statement is the object of:

Lemma 5.6: If A is a Fredholm operator of index zero and 0 is an isolated eigenvalue, then $i(A^{2^n}) = 0$.

Proof: $A^{2^n} - \lambda I$ is also a Fredholm operator for $|\lambda|$ small, of index $i(A^{2^n})$, moreover $n(A^{2^n} - \lambda I)$ and $d(A^{2^n} - \lambda I)$ are constants in a punctured neighborhood of 0 (Goldberg [14], Cor. V.1.7). So it is sufficient to prove that $n(A^{2^n} - \lambda I) = d(A^{2^n} - \lambda I) = 0$ for real positive λ.

The proof is by induction: For $n = 0$ this is the statement of (iii). Suppose the lemma is true for n.
Then $A^{2^{n+1}} - \lambda I = (A^{2^n} - \sqrt{\lambda}\, I)(A^{2^n} + \sqrt{\lambda}\, I)$ on $D(A^{2^{n+1}})$.

By induction hypothesis $A^{2^n} \pm \sqrt{\lambda}\, I$ is a one to one onto map from $D(A^{2^n})$ onto E.

Let x in $D(A^{2^{n+1}})$ be such that $(A^{2^{n+1}} - \lambda I)x = 0$. So $(A^{2^n} + \sqrt{\lambda})x \in D(A^{2^n}) \cap \ker(A^{2^n} - \sqrt{\lambda}\, I) = \{0\}$ and since x is in $D(A^{2^n})$: $x = 0$. Similarly if y is in E, there exists x_1 in $D(A^{2^n})$ such that: $y = (A^{2^n} - \sqrt{\lambda}\, I)x_1$ and there exists x in $D(A^{2^n})$ such that $x_1 = (A^{2^n} + \sqrt{\lambda}\, I)x$. This shows that x belongs to $D(A^{2^{n+1}})$ and $(A^{2^{n+1}} - \lambda I)x = y$. This ends the proof of theorem 5.1.

Example 5.7: Before going to more general perturbation terms, a simple example of a non-isolated eigenvalue will show the

possibility of non-existence of bifurcation: (Taylor [37], p.32).
Let $E = \ell^2$, an element in E is
$x = \{x_1, x_2, \ldots,\}$ such that $\Sigma|x_i|^2 < \infty$. Let $B = \{\{0, x_2, x_3, \ldots\}\}$.
$A(\{0, x_2, x_3, \ldots\}) = \{0, x_4, x_3, x_6, x_5, \ldots\} = y$ with
$y_{2k} = x_{2k+2}, y_{2k+1} = x_{2k+1}$. If B is given the induced ℓ^2 norm
A is continuous with norm 1. $\ker A$ is generated by $\{0,1,0,\ldots\}$
defining P and $R(A)$ has codimension one, with $I - Q$ being
the projection on the first coordinate. Then
$K(\{y_1, y_2, \ldots\}) = \{0, 0, y_3, y_2, \ldots, y_{2k+1}, y_{2k}, \ldots\}$.
$Ax - \lambda x = 0 = \{0, x_4 - \lambda x_2, (1-\lambda)x_3, \ldots, x_{2k+2} - \lambda x_{2k}, (1-\lambda)x_{2k+1}, \ldots\}$
is solvable for $|\lambda| < 1$ and an eigenvector can be represented
by: $\{0, 1, 0, \lambda, \ldots, 0, \lambda^k, 0, \ldots\}$. It is easy to check that
$(I-Q)(I-\lambda KQ)^{-1}P \equiv 0$ and that $\alpha(A) \equiv \infty$. Let
$g(x,\lambda) = \{r(x,\lambda), 0, \ldots\}$ so $Qg(x,\lambda) = 0$. Then (7) and (8)
reduce to:

$$(I-P)x - (I-\lambda KQ)^{-1}\lambda KQPx = 0 \tag{7}$$

$$(I-Q)g(x,\lambda) = r(x,\lambda) = 0. \tag{8}$$

So if, for example, $r(x,\lambda) = ||Px||^2$, there is no non-trivial
solution to $Ax - \lambda x = g(x,\lambda)$.

§ VI : BIFURCATION THEORY FOR $Ax - T(\lambda)x = g(x,\lambda)$.

This section is devoted to the study of (2) in its generality, in order to get closer to the Taylor expansion motivation of the introduction to this chapter.
Here A is a Fredholm operator of non-negative index and defined on a Banach space B with range in another Banach space E.
$T(\lambda) = \lambda T_1 + \ldots + \lambda^p T_p + \ldots$ where T_i are bounded operators

from B to E, such that for small λ the series is uniformly convergent in norm.

$g(x,\lambda)$ satisfies the conditions (1.6).

6.1: Reduction to a finite dimensional system.

Let P be a projection from B onto $\ker A$ and Q be a projection from E onto $R(A)$. As before, by the open mapping theorem, there exists an operator K from E to B such that:

$$AKQ = Q \quad \text{and} \quad KA(I-P) = I - P.$$

Let x in B be written as: $x = x_1 + x_2$ with $x_1 = Px$.

Decompose (2) on E: $\begin{cases} Ax_2 - QT(\lambda)(x_1+x_2) = Qg(x_1+x_2,\lambda) \\ (I-Q)T(\lambda)(x_1+x_2) + (I-Q)g(x_1+x_2,\lambda) = 0 \end{cases}$

Apply K to the first equation, and since $||KQT(\lambda)|| < 1$ for $|\lambda|$ small enough:

$$(10) \quad x_2 - (I-KQT(\lambda))^{-1} KQT(\lambda)x_1 - (I-KQT(\lambda))^{-1} KQg(x_1+x_2,\lambda) = 0$$

This equation is uniquely solvable, by contraction as before, for $x_2 = x_2(x_1,\lambda)$ as a continuous function of x_1 and λ, with estimates similar to those of equation (7).

Then, the second equation becomes:

$$(I-Q)T(\lambda)(I+(I-KQT(\lambda))^{-1}KQT(\lambda))x_1+(I-Q)(I+T(\lambda)(I-KQT(\lambda))^{-1}KQ)g(x,\lambda)$$

Since $I + (I-KQT(\lambda))^{-1} KQT(\lambda) = (I-KQT(\lambda))^{-1}$

and $I + T(\lambda)(I-KQT(\lambda))^{-1} KQ = (I-T(\lambda)KQ)^{-1}$

one has the bifurcation equation:

$$(11) \quad (I-Q)T(\lambda)(I-KQT(\lambda))^{-1}x_1+(I-Q)(I-T(\lambda)KQ)^{-1}g(x_1+x_2(x_1,\lambda),\lambda)=0$$

Denote $(I-Q)T(\lambda)(I-KQT(\lambda))^{-1}\Big|_{\ker A}$ by $B(\lambda)$ so that, given bases for $\ker A$ and coker A, $B(\lambda)$ is a $d^* \times d$ matrix with entries which are analytic in λ as in section IV.

Before deriving the analogues of theorems 3.2, 3.4, and 4.4, and extension of the notion of isolated eigenvalue is needed.

Lemma 6.2: Given A and $T(\lambda)$ as above, there are two non-negative integers n_1 and n_2, and a positive number λ_o, such that for all λ, $0 < |\lambda| < \lambda_o$ then:

i) $A - T(\lambda)$ is a Fredholm operator of index $i(A-T(\lambda)) = i(A)$.

ii) $n(A-T(\lambda)) = n_1 \leq n(A)$

iii) $d(A-T(\lambda)) = n_2 \leq d(A)$

Proof: i) is the content of Goldberg [14] theorem V.1.6., for $||T(\lambda)|| < \gamma(A)$ where $\gamma(A)$ is the minimum modules of A. $\gamma(A)$ is not zero since A is a closed operator (Goldberg [14] Theorem IV.1.6). So, since $||T(\lambda)|| < |\lambda|\ a(\lambda)$ where $a(\lambda) \geq 0$ is a continuous function of λ in a disc of radius ρ around the origin and if $a_o = \sup_{|\lambda| \leq \rho} a(\lambda)$, then i) is true for:

$$|\lambda| < \min(\rho, \frac{\gamma(A)}{a_o}).$$

To prove ii) and iii) it is then enough to show that $n(A-T(\lambda)) = n_1 \leq n(A)$. Take $\lambda_1 = \min\left(\frac{1}{2}\rho, \frac{\gamma(A)}{2\ a_o}, \frac{1}{2||KQ||a_o}\right)$ then the reduction (10), (11) is valid and $(A-T(\lambda))x = 0$ is equivalent to:

(10) $$x_2 - (I-KQT(\lambda))^{-1} KQT(\lambda)x_1 = 0$$

(11) $$B(\lambda)x_1 = 0$$

Given x_1, the first equation is uniquely solvable for x_2, so $n(A-T(\lambda)) = \dim\ker B(\lambda) = n(A) - \text{rank } B(\lambda)$.

Consider all possible minors of $B(\lambda)$; each has a determinant which is analytic in λ, vanishes at 0 and, unless identically zero, cannot accumulate zeros at 0. So let $\lambda_o = \min(\lambda_1, \tilde{\lambda})$ where $\tilde{\lambda}$ is chosen such that for all λ, with $0 < |\lambda| < \tilde{\lambda}$, all the non-trivial determinants do not vanish in this range. Let $\tilde{B}(\lambda)$ be a minor of $B(\lambda)$ of maximal dimension $n(A) - n_1$ and whose determinant is not trivial, i.e. for $0 < |\lambda| < \lambda_o$, determinant of $\tilde{B}(\lambda)$ is never zero. Then rank $B(\lambda)$ is $n(A) - n_1$ and so proving the lemma by choice of $\tilde{B}(\lambda)$. (Note that if all minors are identically zero then: $n(A-\lambda I) = n(A)$).

<u>Lemma 6.3</u>: Assume $T(\lambda)$ is analytic for all λ in $\mathbb{K}$. Let U be the set of λ in $\mathbb{K}$ such that: $A - T(\lambda)$ is a Fredholm operator. Then:

i) U is open.

ii) If $\mathcal{E}$ is a connected component of U, then there are constant integers n_o, n_1, n_2 such that: $i(A-T(\lambda)) = n_o$ for all λ in $\mathcal{E}$, $n(A-T(\lambda)) = n_1$ and $d(A-T(\lambda)) = n_2$ in $\mathcal{E}$ except at isolated points λ_i where $n(A-T(\lambda_i)) > n_1$ and $d(A-T(\lambda_i)) > n_2$.

<u>Proof</u>: i) If λ_o is in U, then $A - T(\lambda) = A - T(\lambda_o) - (T(\lambda)-T(\lambda_o))$ is a Fredholm operator for $|\lambda-\lambda_o|$ small enough.

Let $n_1 = \min_{\mathcal{E}} n(A-T(\lambda))$ and suppose n_1 is obtained at λ_o.

Since U is open, $\mathcal{E}$ is open in $\mathbb{K}$ and hence path connected. So if $\tilde{\lambda}_o$ is a point in $\mathcal{E}$ such that $n(A-T(\tilde{\lambda}_o)) > n_1$, there is a path from λ_o to $\tilde{\lambda}_o$ which may be covered by a finite number of disks $S(\lambda_i)$ contained in $\mathcal{E}$ and with centers at $\lambda_o, \lambda_1, \lambda_2, \ldots, \lambda_k = \tilde{\lambda}_o$ with the properties listed in lemma 6.2. On $S(\lambda_o)$: $n(A-T(\lambda)) = n_1$, since by lemma 6.2 $n(A-T(\lambda)) \leq n_1$ the equality being obtained from the choice of n_1. So $n(A-T(\lambda)) = n_1$ on $S(\lambda_o) \cap S(\lambda_1)$ and hence on $S(\lambda_1) - \{\lambda_1\}$. Similarly $i(A-T(\lambda)) = i(A-T(\lambda_o))$ on $S(\lambda_1)$ and so on ... So $n(A-T(\lambda)) = n_1$ on the path except possibly at $\lambda_1, \lambda_2, \ldots, \lambda_k$ but the index remains constant.

<u>Assume</u>:

(12) $\quad i(A) \geq 0$ and $d(A-T(\lambda)) = 0$ for $0 < |\lambda| < \lambda_o$

Then $B(\lambda)$ has maximal rank and, given a non-trivial $d^* \times d^*$ minor $\tilde{B}(\lambda)$, one gets $\det \tilde{B}(\lambda) = \lambda^m a(\lambda)$ with $a(0) \neq 0$. m will be called the <u>algebraic multiplicity</u> of A at 0.

<u>Theorem 6.4</u>: Assume $i(A) = 0$.

If $\mathbb{K} = \mathbb{R}$, <u>the parity of</u> m <u>and the sign of</u> $a(0)$ <u>are independent of</u> P <u>and</u> Q.

If $\mathbb{K} = \mathbb{C}$, m <u>is independent of</u> P <u>and</u> Q.

<u>Proof</u>: It is enough to note that in the proof of theorem 4.5, only the analyticity of $\det \tilde{B}(\lambda)$ was used.

<u>Theorem 6.5</u>: <u>Suppose that (12) holds, and that</u> g <u>satisfies (1.6) with</u> $n = m - d^* + 1$. <u>Then the conclusions of theorem 4.4 are valid</u>, i.e.:

i) If $i(A) = 0$ *and* m *is odd, bifurcation takes place*.

ii) *Assume* $i(A) = 0$ *and* $n(A) = 1$, *then*: (1) *If* $K = R$ *and* m *is odd, there are two bifurcating branches*.
(2) If $\mathbb{K} = \mathbb{C}$ *and any* m, *there is bifurcation in all directions in* $\ker A$.

iii) If $i(A) > 0$: *there is an* $i(A)$ - *parameter family of bifurcating solutions*.

Proof: Recall that the proofs of theorems 4.4, 3.2 and 3.4 did not use the special form of $T(\lambda)$.

CHAPTER TWO

BIFURCATION THEORY FOR SEVERAL PARAMETERS

The viewpoint of Section VI in the last chapter leads in a natural way to the study of bifurcation problems involving more than one parameter. i.e. for

(2) : $$Ax - T(\lambda)x = g(x,\lambda)$$

where A is a Fredholm operator of non-negative index from the real or complex Banach space B to the real or complex Banach space E.

$T(\lambda)$, where λ denotes the vector $(\lambda_1,\ldots,\lambda_n)$, is an operator analytic in λ, i.e.:

$$T(\lambda) = \Sigma\lambda_1^{i_1}\ldots\lambda_n^{i_n} T_{i_1\ldots i_n} \quad \text{with} \quad \sum_{j=1}^{n} i_j = k, k=1,\ldots$$

and $T_{i_1\ldots i_n}$ are bounded operators and the series is uniformly convergent for small λ.

Furthermore $g(x,\lambda)$ satisfies condition (1.6) of chapter one, where $|\lambda-\lambda'|$ is replaced by the euclidean norm in $\mathbb{K}^n$.

It is clear that the reduction (10), (11) is still valid, with the same estimates, i.e. (2) is equivalent to:

(10) $$x_2 - (I-KQT(\lambda))^{-1}KQT(\lambda)x_1 - (I-KQT(\lambda))^{-1}KQg(x_1+x_2,\lambda) = 0$$

which is uniquely solvable by contraction, for $x_2 = x_2(x_1,\lambda)$ with:

$$||x_2(x_1,\lambda)-x_2(x_1',\lambda')||_B \leq \tilde{C}(M(r)||x_1-x_1'||_B+||\lambda-\lambda'||+||\lambda||\ ||x_1-x_1'||_B)$$

and

$$||x_2(x_1,\lambda)||_B \leq \tilde{C}(||x_1||_B^2 + ||\lambda||^{2n'+1} + ||\lambda||\ ||x_1||_B)$$

for

$$||x_1||_B,\ ||x_1'||_B \leq r < r_o, ||\lambda||\ ,\ ||\lambda'|| < \lambda_o$$

where the terms $||\lambda||\ ||x_1 - x_1'||_B$ and $||\lambda||\ ||x_1||_B$ came from $(I-KQT(\lambda))^{-1}KQT(\lambda)x_1$ and n' will be defined later.

And

$$(11) \quad \begin{cases} (I-Q)T(\lambda)(I-KQT(\lambda))^{-1}\ x_1 \\ \\ + (I-Q)(I-T(\lambda)KQ)^{-1}\ g(x_1+x_2(x_1,\lambda),\lambda) = 0 \end{cases}$$

which is a finite dimensional bifurcation equation, from a neighborhood of $(0,0)$ in $\mathbb{K}^d \times \mathbb{K}^n$ into $\mathbb{K}^{d*}$ where, given bases for ker A and coker A, $(I-Q)T(\lambda)(I-KQT(\lambda))^{-1}\Big|_{\ker A} = B(\lambda)$ is an analytic function of the n variables $\lambda_1,\ldots,\lambda_n$.

The aim of this chapter is to study equation (11) without assuming any special form for the nonlinearity, except smallness and regularity, in contrast with Chapter four where such special assumptions on g will be made.

Thus, in order to apply topological methods, one needs to deform (11) to its known part, i.e. $B(\lambda)x_1$, and so, it is necessary that on the boundary of some ball, $B(\lambda)x_1$ and maybe some side condition as $||x_1||^2 - r^2$ or fixing $x_1 \neq 0$, is different from zero and dominates the nonlinearity.

Remark 1: Note that one should not be restricted to spheres and that in principle any set, on which $B(\lambda)x_1$ and side condition are not zero, could give interesting results. This leads naturally to the study of homotopy classes of maps from some set in $\mathbb{K}^d \times \mathbb{K}^n$ into $\mathbb{K}^{d*} - \{0\}$, i.e.to cohomotopy groups and the extension of the notion of essential maps. This approach will be used in the next chapter but here, for the existence of solutions in the small, balls and spheres seem more suitable and certainly much easier to handle.

So, one hopes that these d^* analytic functions of $d + n$ unknowns, (or less, if one specifies a certain number of them), have an isolated common zero at 0 and then one should try to compute the homotopy type of the resulting mapping. The first point is an important part of algebraic geometry, the second of homotopy theory. So, it is clear that this program is too ambitious to carry out in its full generality, but one can try to obtain results which are easy to use, by approaching this study along the lines of theorems 3.2 and 3.4 of Chapter I. That is, assuming that A has index zero, see if $B(\lambda)$ is non-singular for certain regions of the λ space, or if for some choice of x_1 one has a non-trivial mapping in the λ space only. (Or use a mixing of these two approaches.)

<u>Remark 2</u>: In the complex case, for A of index zero, $B(\lambda)$ cannot be non-singular in a punctured neighborhood of 0 except if $n = 1$. In fact, $\det B(\lambda)$ is a holomorphic function, f, from $\mathbb{C}^n$ to $\mathbb{C}$ and, if not identically zero, the Weierstrass preparation theorem permits to write the germ of f as: $[f] = UP$ where U is a unit and P a polynomial in one of the λ's. P in turn can be decomposed in a product of irreducible Weierstrass polynomials P_i. Then the germ of zeros of f form a variety V which decomposes, by the Nullstellensatz (Gunning-Rossi [18] theorem 19, p. 91, Fulton [12], p. 22) into irreducible branches,

corresponding to P_i, each having local dimension $n - 1$ at the point 0. (Gunning-Rossi [18] theorem 11, p. 113). So for $n > 1$, $\det B(\lambda)$ cannot have an isolated zero.

Note that this chapter is a collection of simple facts. Theorems give special emphasis on the most important and useful among these.

§ I : A SPECIAL CASE: $d = d^* \leq n$

This case is the simplest, from the topological point of view, and is usually the only one considered in numerical analysis, where one tries to expand all the unknowns in power series of a single parameter. This section is an extension of theorem 3.2 of the last chapter and only degree theory will be used.

Remark 1.1: A similar use of degree theory for several parameters is given in [3].

Now, if $x_1 = (z_1, \dots, z_d)$, (11) has the form:

$$\begin{cases} a_{11}(\lambda)z_1 + \dots + a_{1d}(\lambda)z_d + g_1(z_1, \dots, z_d, \lambda) = 0 = P_1(x_1, \lambda) + g_1(x_1, \lambda) \\ \dots\dots\dots\dots\dots\dots \qquad \dots\dots\dots\dots \qquad \dots\dots\dots\dots\dots\dots \\ a_{d1}(\lambda)z_1 + \dots + a_{dd}(\lambda)z_d + g_d(z_1, \dots, z_d, \lambda) = 0 = P_d(x_1, \lambda) + g_d(x_1, \lambda) \end{cases}$$

where, from the estimate for $x_2(x_1, \lambda)$:

$$||g|| \leq C(||x_1||^2 + ||\lambda||^{2n'+1}).$$

Suppose that for some x_1^o and $\lambda_{i_1}, \dots, \lambda_{i_{n-d}}$ fixed, with

$$||x_1^o|| = r \text{ and } (\lambda_1, \dots, \lambda_n) = (\lambda_1, \dots, \lambda_d, \lambda_{i_1}, \dots, \lambda_{i_{n-d}}),$$

$P_i(x_1^o, \lambda)$ have an isolated common zero at $\lambda_1 = \dots = \lambda_d = 0$.

Furthermore, assume that there exist ρ such that, for $(\sum_1^d |\lambda_i|^2)^{1/2} = \rho$, then $||P|| > ||g||$, so that the degree of the map defined by (11) on the ball $(\sum_1^d |\lambda_i|^2)^{1/2} \leq \rho$ is well defined and equal to the index of the field defined by $(P_1, \ldots, P_d) = P$. If this index is not zero, then there is a solution to (11) with $||x_1^o|| = r$.

A particular case of this is the following:

<u>Suppose</u> $n = d$ and $P_i(x_i^o, \lambda) = P_i^{k_i}(x_1^o, \lambda) + Q_i^{k_i+1}(x_1^o, \lambda)$ where $P_i^{k_i}(x_1^o, \lambda)$ are homogeneous polynomials in $\lambda_1, \ldots, \lambda_d$, of degree k_i, $Q_i^{k_i+1}(x_1^o, \lambda)$ are polynomials beginning with terms of order $k_i + 1$. Furthermore, assume that $P_i^{k_i}(x_1^o, \lambda)$ have an isolated common zero at $\lambda_1 = \ldots = \lambda_d = 0$.

<u>Theorem 1.2</u>: <u>Suppose</u> $|g_i| \leq c(||x_1||^2 + ||\lambda||^{\alpha+k_i+1})$ <u>where</u> $\alpha = \max k_i$ $i = 1, \ldots, d$. <u>Then</u>:

i) If $\mathbb{K} = \mathbb{R}$ <u>and all</u> k_i <u>are odd: there is bifurcation</u>.

ii) If $\mathbb{K} = \mathbb{C}$ <u>there is always bifurcation</u>.

<u>Proof</u>: Noting that $P_i(tx_1^o, \lambda)$ and $P_i^{k_i}(tx_1^o, \lambda)$ have the same properties of non-degeneracy for all t, $0 < t \leq 1$, one can assume that $||x_1^o|| = r$ is arbitrarily small. Moreover for each λ_o, $\lambda_o \neq 0$, there exists an i such that:

$P^{k_i}(x_1^o, \lambda_o) \neq 0$ and so by homogeneity:

$$|P_i^{k_i}(x_1^o, \lambda_o)| > ||x_1^o|| \, A\left(\frac{\lambda_o}{||\lambda_o||}\right) ||\lambda_o||^{k_i}$$

and

$$|Q_i^{k_i+1}(x_1^o,\lambda_o)| \le ||x_1^o||B(\frac{\lambda_o}{||\lambda_o||})||\lambda_o||^{k_i+1}.$$

These estimates still hold for a neighborhood of $\lambda_o/||\lambda_o||$ in the unit sphere and hence, by compactness, one can choose A and B such that for each $\lambda_o, \lambda_o \neq 0$, there is an i with:

$$\left.\begin{aligned} &|P_i^i(x_1^o,\lambda_o)| > ||x_1^o||\ A\ ||\lambda_o||^{k_i} \\ &\text{and} \\ &|Q^{k_i+1}(x_1^o,\lambda)| \le ||x_1^o||\ B\ \ ||\lambda_o||^{k_i+1}. \end{aligned}\right\}$$

So if $||\lambda_o|| = (Mr)^{1/\alpha}$, $||x_1^o|| = r$ then:

$$Ar(Mr)^{k_i/\alpha} > Br(Mr)^{(k_i+1)/\alpha} + C(r^2+(Mr)^{(\alpha+k_i+1)/\alpha})$$

for M so large that: $M > 1$ and $A - CM^{-k_i/d} \ge \frac{A}{2}$

and $r \le (A^\alpha 2^{-\alpha} M^{-1}(B+MC)^{-\alpha}$, 1, radius of convergence).

The theorem follows, once one recalls that the index of the vector field $P_i^{k_i}$ is $\prod_{i=1}^{d} k_i$ if $\mathbb{K} = \mathbb{C}$ and $\prod_{i=1}^{d} k_i \bmod[2]$ if $\mathbb{K} = \mathbb{R}$, Cronin [6] p. 45-49. Sather [30] p.223 and 234.

Remark 1.3: Similar results are valid when $n > d$, but the estimates on g are more complicated to formulate.

Remark 1.4: In the real case, the index may be not zero when some k_i is even as the study for $d = 2$, $n = 2$ will show.

Remark 1.5: If $\det(B(\lambda)) \neq 0$ for all $\lambda, \lambda \neq 0$, then under suitable smallness conditions for g and if the index of $P_i(x,\lambda)$

is different from zero for some x_1^o, then bifurcation occurs in all directions in ker A (since any x_1 in ker A can be reached by a rotation from a fixed vector of same norm, so the index will remain constant). Note that, if $n > d$, this index is always zero (deformation through the remaining λ's) and that $B(\lambda)$ non-singular, implies $\mathbb{K} = \mathbb{R}$, or $n = 1$ in the complex case.

As an example, the case $\mathbb{K} = \mathbb{R}$, $d = n = 2$, $\det B(\lambda) > 0$ for $\lambda \neq 0$ will be treated. Then (11) reduces to:

$$\begin{bmatrix} a(\lambda) & b(\lambda) \\ c(\lambda) & d(\lambda) \end{bmatrix} \begin{bmatrix} z_1 \\ z_2 \end{bmatrix} + \begin{bmatrix} g_1(z,\lambda) \\ g_2(z,\lambda) \end{bmatrix} = 0$$

and $\det B(\lambda) = P_{k_o}(\lambda_1,\lambda_2) + P_{k_o+1}(\lambda_1,\lambda_2) + \ldots$

where $P_{k_o}(\lambda_1,\lambda_2)$ is a homogeneous polynomial of degree k_o.

$P_{k_o}(\lambda_1,\lambda_2) = \prod_{i=1}^{s_o} (a_i\lambda_1 - b_i\lambda_2)^{p_i} N_o(\lambda_1,\lambda_2)$ where $N_o(\lambda_1,\lambda_2)$

is a homogeneous polynomial of even degree, vanishing only at 0 and otherwise positive. It is easy to check that k_o is even, p_i are even and that for each direction $a_i\lambda_1 - b_i\lambda_2 = 0$ there exists a first non-zero term P_{k_i}, with k_i even, $P_{k_i}(\frac{b_i}{a_i}\lambda_2,\lambda_2)$ is positive except for $\lambda_2 = 0$, and that $P_{k_o}(\lambda_1,\lambda_2) \geq 0$. Finally, it can be shown that there exist an integer α and a constant $A > 0$ such that, if $\lambda_1^2 + \lambda_2^2 = \rho^2 \leq \rho_o^2$, then $\det B(\lambda) \geq A\rho^\alpha$ (Krasnosel'skii [22]' Appendix). Note that $\alpha \geq k_i$ for all i, and that $P_{k_o} + \ldots + P_\alpha \geq \frac{A}{2}\rho^\alpha$.

With respect to g suppose that $||g|| \leq C(||x||^2 + \rho^{2\alpha-1})$.

Then it is easy to check that, since $||B(\lambda)^{-1}|| \leq D^{-1}\rho^{1-\alpha}$ for ρ small enough, the right hand side, for $z_1^2 + z_2^2 = r^2$, has norm $||g|| \leq C(r^2 + \rho^{2\alpha-1})$, while the left $||B(\lambda)x_1|| \geq Dr\,\rho^{\alpha-1}$. If $\rho = (Mr)^{1/\alpha-1}$ as before: $Dr(Mr) > C(r^2+(Mr)^2(Mr)^{1/\alpha-1})$ for M large enough and r small enough. So one can deform (11), for a fixed $z_1,z_2(z_1^2+z_2^2=r^2)$ on the sphere $\rho = (Mr)^{1/\alpha-1}$, to $(a(\lambda)z_1+b(\lambda)z_2, c(\lambda)z_1+d(\lambda)z_2)$ and even to polynomials, by discarding any term of homogeneous degree higher than α; the degree of this map, by remark 1.5, will be the index of $(a(\lambda)r, c(\lambda)r)$ (one may take $r = 1$). Suppose that $a(\lambda) = P_{k_1}(\lambda_1,\lambda_2) + \tilde{d}(\lambda)$, $c(\lambda) = Q_{k_2}(\lambda_1,\lambda_2) + \ldots$ then, if P_{k_1} and Q_{k_2} have no common factor and k_1k_2 is odd, this index, by theorem 1.2, is odd.

Example 1.6: $k_1 = 1$ so $P_{k_1}(\lambda_1,\lambda_2) = a\lambda_1 + b\lambda_2 = \mu$.

If one has $Q_{k_2}(\lambda_1,\lambda_2) = \mu Q_{k_2-1}(\lambda_1,\lambda_2)$ then, by using the deformation: $(a(\lambda), c(\lambda)-tQ_{k_2-1}(\lambda)a(\lambda))$, one gets

$(\mu+\tilde{d}(\lambda), Q_{k_2+1} - Q_{k_2-1}\tilde{d}(\lambda) + \ldots)$ that is $c(\lambda)$ has been replaced by a series with a starting polynomial of homogeneous degree at least $k_2 + 1$. Then, if μ divides this polynomial, one can repeat the deformation until one gets: (det $B(\lambda)$ is non-degenerate) $(\mu+\tilde{d}(\lambda), Q_m+e(\lambda))$ where μ and Q_m are prime, $e(\lambda)$ and $\tilde{d}(\lambda)$ are of higher order. As in theorem 1.2, one can reduce this vector field to (μ,Q_m).

Lemma 1.7: (μ,Q_m) has an index 0 if m is even, ± 1 if m is odd.

Proof: Since μ and Q_m are prime, $Q_m = a\,\nu^m + \mu\,Q_{m-1}$ with $a \neq 0$, ν and μ are linearly independent combinations of

λ_1 and λ_2. Then $(\mu, a\nu^m + (1-t)\mu Q_{m-1})$ is an allowable deformation and, if m is even $m = 2p$, $(\mu(1-t), a((1-t)\nu^{2p}+t))$ gives a constant map, while if m is odd, $m = 2p + 1$, $(\mu, a((1-t)\nu^{2p} + t)\nu)$ has degree, in the λ_1, λ_2 plane, sign $a \times$ (sign of the transformation: $\lambda_1, \lambda_2 \leftrightarrow \mu, \nu$).

<u>Remark 1.8</u>: When one of the k's is even, it is still possible to have a non-zero degree. For example $(\lambda_1^2-\lambda_2^2, \lambda_1\lambda_2)$ has index 2.

Many other examples and means of computing the index of plane vector fields can be found in [22]'.

§ II : GENERAL SITUATION

As pointed out in the introduction to this chapter, one tries to compute some topological invariant for equation (11), which will give the existence of bifurcation; in order to be able to do so, $B(\lambda)x_1$ has to dominate the nonlinear part of (11) in a certain region of the λ space. In the complex case, as it was seen in Remark 2, $\det B(\lambda) = 0$ is a variety V whose irreducible components have local dimension $n - 1$. And in fact, if one looks for solutions (x_1, λ) with λ belonging to a subvariety W of $\mathbb{C}^n$ of dimension k, then $\dim(V \cap W) \geq k - 1$ (Gunning-Rossi [18], Theorem 14, p.115). So, there is no hope to deform the nonlinearity or to define an invariant from some sphere (in $\mathbb{C}^k \times \mathbb{C}^d$) to another sphere (in $\mathbb{C}^{d^*}$) unless k is one. This means that one is forced to reduce the n-parameter problem to a one parameter problem.

So, suppose that $\det B(\lambda) = P_{k_o}(\lambda_1, \ldots, \lambda_n) +$

$+ P_{k_o+1}(\lambda_1,\ldots,\lambda_n) + \ldots$ with $P_{k_o} \not\equiv 0$, P_i are homogeneous polynomials of degree i, $P_{k_o}(\lambda_1,\ldots,\lambda_n) = 0$ forming a real or complex projective variety of dimension at most $n - 1$. For the case of positive index, assume $B(\lambda)$ represents a $d^* \times d^*$ minor whose determinant is not identically zero. (This replaces conditions (9) and (12) of the first chapter).

Theorem 2.1: *Suppose*: $g(x,\lambda)$ *satisfies condition (1.6) of chapter one with* n *replaced by* $k_o - d^* + 1$. *Then*:

i) $i(A) = 0$, *if* k_o *is odd there exists an* $(n-1)$-*parameter family of bifurcating branches.*

ii) $i(A) > 0$, *there exists an* $(n-1 + i(A))$-*parameter family of bifurcating branches.*

Proof: Choose a direction λ_o, $||\lambda_o|| = 1$, such that $P_{k_o}(\lambda_o) \neq 0$. Without loss of generality, one can assume that this direction corresponds to one of the axis, say λ_1, i.e. $\lambda_2 = \ldots = \lambda_n = 0$. Then (11) is reduced to one parameter λ_1, and one can apply theorem 6.5 of chapter I, for the direction λ_o and in fact for any other direction close to λ_o.

Theorem 2.2: *If* k_o *is even, suppose that there is a direction* λ_o *such that* $P_{k_o}(\lambda_o) = \ldots = P_{k_o+m-1}(\lambda_o) = 0$ *and* $P_{k_o+m}(\lambda_o) \neq 0$ *with* m *odd. Then, for the appropriate estimate on* g (i.e. $n = k_o + m - d^* + 1$) *bifurcation takes place in the* λ_o *direction.*

The proof is similar to the proof of theorem 2.1.

Remark 2.3: When this approach fails it is possible to seek

solutions (x,λ) with λ on some variety V of higher dimension (this applies only to the real case) and, if $\det B(\lambda) \neq 0$ on $V-\{0\}$, study the homotopy of $B(\lambda)x_1$ and a side condition as $||x_1||^2 - r^2$ on a sphere in $\ker A \times V$ surrounding the origin. One has then to use higher homotopy groups of spheres, Toda [35].

Example 2.4: Let $P_{k_o}(\lambda_1,\lambda_2,\lambda_3) = \lambda_1^2 + \lambda_2^2 - \lambda_3^2$. $\mathbb{K} = \mathbb{R}$. Since one cannot apply theorem 2.1, one seeks a solution with $\lambda_3 = \varepsilon\rho$ fixed, $\lambda_1^2 + \lambda_2^2 \leq \rho^2$ where ρ may be of the form $(Mr)^{1/n}$, i.e. study $(B(\lambda)x_1, ||x_1||^2 - r^2)$ on $\{||x_1||^2 + \lambda_1^2 + \lambda_2^2 = r^2 + (Mr)^{2/n}\}$ with the familiar condition on g. This gives a map from S^{d+1} to S^d and, if its homotopy class is non-trivial, a one-parameter family of solutions (ε varying). For instance if

$$B(\lambda)x_1 = \begin{bmatrix} \lambda_1-\lambda_3 & \lambda_2 \\ -\lambda_2 & \lambda_1+\lambda_3 \end{bmatrix}\begin{bmatrix} z_1 \\ z_2 \end{bmatrix} \quad \text{then}$$

for $\lambda_3 = 0$, $B(\lambda)\, x_1$ and $||x_1||^2 - r^2$ is the Hopf map $(\lambda = \lambda_1 + i\lambda_2)$.

Remark 2.5: If $P_{k_o}(\lambda)$ is not degenerate (i.e. vanishing only at 0), any attempt to get a solution on a subvariety of $\mathbb{R}^n$ will fail and one has to look for a map from S^{d+n-1} to S^d. The systematic study of this situation, with the aid of more sophisticated topological tools, will be the object of a forthcoming article.

Remark 2.6: It is possible to prove the analogue of theorem 6.4

in chapter one, i.e. that the parity of k_o or k_o itself, is independent of the choices of P, Q, K, since the proof relies on the fact that 0 is an isolated point, in certain λ directions, of the $T(\lambda)$ spectrum of A.

<u>Remark 2.7</u>: If there are polynomials $P_1(\lambda),\ldots,P_n(\lambda)$ with an isolated zero at 0, such that $B(\lambda_1,\ldots,\lambda_n) = \tilde{B}(P_1(\lambda),\ldots,P_n(\lambda))$ then the homotopy class of $B(\lambda)x_1$ with any side condition will be equal, by functoriality, to the homotopy class of $\tilde{B}(\mu_1,\ldots,\mu_n)x_1$ and side condition multiplied by the index of the vector field $(\lambda_1,\ldots,\lambda_n) \to (P_1(\lambda),\ldots,P_n(\lambda))$. For example if $T(\lambda) = \lambda_1 T_1 + \lambda_2^3 T_2$ the computations will be easier if one replaces λ_2^3 by μ_2.
Note also that if $g(x,\lambda)$ depends on λ via $P_1(\lambda),\ldots,P_n(\lambda)$, then one can replace λ by μ in the original equation and, if $\tilde{B}(\mu)x_1$ and side condition have a non-trivial homotopy class, one will get then a solution, even if the index of the transformation $\lambda \to \mu$ has zero index; (In this case one may have more than one solution).

In order to see how to use the preceding remarks, the case of two parameters will be studied, for $\mathbb{K} = \mathbb{R}$ and $\det B(\lambda) > 0$ except at 0. Then one has to use a mapping from S^{d+1} to S^d. Note that if $d = 1$ any such map is trivial, since $\pi_m(S^1) = 0$ for all $m > 1$ (Spanier [34]).

<u>Remark 2.8</u>: If $n = 2$, $d = 2$, $B(\lambda) = \begin{bmatrix} a(\lambda) & b(\lambda) \\ c(\lambda) & d(\lambda) \end{bmatrix}$

Then $\begin{bmatrix} a(\lambda) & (1-t)b(\lambda)-tc(\lambda) \\ c(\lambda) & (1-t)d(\lambda)+ta(\lambda) \end{bmatrix}$ has a determinant:

$(1-t) \det B(\lambda) + t(a(\lambda)^2 + c(\lambda)^2)$ which is positive except at 0. So $B(\lambda)$ is reduced to $\begin{bmatrix} a(\lambda) & -c(\lambda) \\ c(\lambda) & a(\lambda) \end{bmatrix}$ and, denoting $P(\lambda) = a(\lambda) + ic(\lambda)$, one is tempted to study $(P(\lambda)z, ||z||^2 - r^2)$, using Remark 2.7, decomposing it as:

$$(\lambda) \longrightarrow (P(\lambda)) \xrightarrow{\eta} (P(\lambda)z, ||z||^2 - r^2)$$

so its homotopy class is the degree of the first multiplied by the class of the Hopf map. But this is exactly the approach taken in section I so no new result can be expected.

Example 2.9: $n = 2$, $d = 3$, $\det B(\lambda) > 0$ except at 0. One could try to reduce $B(\lambda)$ to a simpler map, by expanding $\det B(\lambda) = a_{11}\,\Delta_1 + a_{12}\,\Delta_2 + a_{13}\,\Delta_3$, where a_{1i} are the elements of the first row and Δ_i the corresponding minors. Then $B(\lambda)x_1$ and $||x_1||^2 - r^2$, on the sphere $\{||x_1||^2 + ||\lambda||^2 = r^2 + (Mr)^{2/n}\}$, has the homotopy type of $(\tilde{B}(\lambda)x_1, ||x_1||^2 - r^2)$ where $\tilde{B}(\lambda)x_1$ is the matrix $B(\lambda)$ with elements in the first row: $(\Delta_1, -\Delta_2, \Delta_3)$. Since the deformation matrix with elements on the first row:

$$((1-t)a_{11} + t\Delta_1,\ (1-t)a_{12} - t\Delta_2,\ (1-t)a_{13} + t\Delta_3)$$

has a non-degenerate determinant, $(1-t) \det B(\lambda) + t(\Delta_1^2 + \Delta_2^2 + \Delta_3^2)$. Then assuming that Δ_1, for instance, is non-degenerate and say positive for $\lambda \neq 0$, one can deform the new matrix to a matrix with elements in the first row $(1,0,0)$, via a deformation matrix of the type $((1-t)\Delta_1 + t,\ -(1-t)\Delta_2,\ (1-t)\Delta_3)$ and lastly deform to zero all the elements of the first column, except the first, without changing the determinant. Then applying the method suggested in Remark 2.8, one can compute the homotopy class of this mapping which is the suspension of a mapping from S^3 to S^2.

For example, $B(\lambda_1,\lambda_2) = \begin{bmatrix} \lambda_2^2 & \lambda_1^2 & \lambda_1\lambda_2^3 \\ -\lambda_1 & \lambda_1 & -\lambda_2 \\ -\lambda_2 & \lambda_2 & \lambda_1 \end{bmatrix}$

with determinant $(\lambda_1^2 + \lambda_2^2)^2$, can be replaced by:

$$\begin{bmatrix} \lambda_1^2+\lambda_2^2 & \lambda_1^2+\lambda_2^2 & 0 \\ -\lambda_1 & \lambda_1 & -\lambda_2 \\ -\lambda_2 & \lambda_2 & \lambda_1 \end{bmatrix} \text{ and deformed to } \begin{bmatrix} 1 & 0 & 0 \\ 0 & \lambda_1 & -\lambda_2 \\ 0 & \lambda_2 & \lambda_1 \end{bmatrix}$$

so that $(B(\lambda)x_1, ||x_1||^2 - r^2)$ has the homotopy type of the suspension of the Hopf map.

Remark 2.10: It is clear that this topological method can be applied to the study of operators A with negative indices, if the number of parameters is enough to assure a mapping between two spheres of same dimension. For example if, for $i(A) = -1$, $n = 2$, $B(\lambda)x_1$ and a side condition define a mapping from a sphere in $\mathbb{R}^d \times \mathbb{R}^2$ into $\mathbb{R}^{d^*} \times \mathbb{R} - \{0\}$ of non-zero degree, one would get a bifurcation point. For instance, if $d = 1$, $d^* = 2$, the equations $(\lambda_1 x, \lambda_2 x)$ for a fixed $x = r$, define a mapping from a sphere in $\mathbb{R}^2$ to $\mathbb{R}^2 - \{0\}$ of degree one.

§ III : APPLICATION: THE HOPF BIFURCATION THEOREM.

In order to illustrate the ideas contained in this chapter, the problem of bifurcation of periodic orbits from an equilibrium point for autonomous differential systems will be considered.

These results can be found in a recent article by J. Alexander and J. Yorke [0], but the methods used by these authors are completely algebraic in nature: Framed bordism, the associated Cech Cohomology theory and the relative Alexander duality between these two theories. The purpose of this section is to give a simple geometrical proof for this important bifurcation result in two parameters while using the methods exposed above.

The problem is the following: study the periodic solutions of the equation:

$$(E_\lambda) : \quad \dot{x} = f(x,\lambda)$$

for x in a n-dimensional C^1 manifold M and λ in some real interval Λ, and where f is a continuous function from $M \times \Lambda$ to TM the tangent bundle of M. It is assumed that, for any x in M and λ in Λ, (E_λ) has a unique solution $G(x,\lambda,t)$ in M starting at x and defined for all t in some open interval of time $(0,a(x,\lambda))$ with $a(x,\lambda) > 0$. Furthermore if x belongs to ∂M, the boundary of M, $G(x,\lambda,t)$ is constrained to remain on ∂M.

Under these conditions, Alexander and Yorke show that one may consider M as an open set in R^n (By first extending the field f to a collar attached to ∂M, then, after embedding M in some euclidian space, by extending f to a tubular neighborhood with a radial component vanishing only on M so that the periodic solutions stay on M).

Then $W = \{(x,\lambda,t) \in M \times \Lambda \times (0,a(x,\lambda))\}$ is an open subset of $M \times \Lambda \times \mathbb{R}^+$ and $G(x,\lambda,t)$ is a continuous map from W to M, ([0] §3).

It is also assumed that for some x_o in M and λ_o in

Λ one has $f(x_o,\lambda) = 0$ for all λ in some open interval Λ_o containing λ_o, so x_o is a stationary point for (E_λ). From now on x_o will be taken to be $x_o = 0$. Finally $f(x,\lambda)$ is supposed to be continuously differentiable in x at $(0,\lambda)$ for λ in Λ_o. That is $f_x(0,\lambda) \equiv L(\lambda)$ is a continuous $n \times n$ matrix such that:

$||f(x,\lambda) - L(\lambda)x|| = o(||x||)$ for λ in Λ_o and x small.

By solving the linearized equation $\dot{x} = L(\lambda)x$, Alexander and Yorke (Section 7) show that $G(x,\lambda,t)$ is continuously differentiable in x at $(0,\lambda,t)$ for λ in Λ_o, t in $\mathbb{R}^+$ with $G_x(0,\lambda,t) = \exp(tL(\lambda))$.

Now it is clear that the study of the periodic solutions of (E_λ) is equivalent to finding the zeros in W of:

$$(3.1) \quad F(x,\lambda,t) = x - G(x,\lambda,t)$$

and (x,λ,t) is stationary only if $f(G(x,\lambda,t),\lambda) = 0$.

Remark 3.2: To study the stationary solutions of (E_λ) near $(0,\lambda_o)$ one has to find the zeros of: $f(x,\lambda) = L(\lambda)x + g(x,\lambda)$ where $||g(x,\lambda)|| = o(||x||)$. This is a classical problem of bifurcation in one parameter. A result in this direction is the following: If $L(\lambda)$ is singular at λ_o and has a determinant which changes sign as λ passes through λ_o then, for all r, $0 < r < r_o$, one has a stationary solution (x,λ) with $||x||=r$.

From now on $L(\lambda_o)$ will be assumed to be non-singular so that the only stationary solutions near $(0,\lambda_o)$ will be of the form $(0,\lambda)$.

Now $F(x,\lambda,t) = (I - \exp tL(\lambda))x + g(x,\lambda,t)$ where

$||g(x,\lambda,t)|| = o(||x||)$ uniformly on bounded sets in $\Lambda \times \mathbb{R}^+$, so if $I - \exp tL(\lambda_o)$ is invertible the only solution of $F(x,\lambda,t) = 0$ near $(0,\lambda_o,t)$ is $(0,\lambda,t)$ which corresponds to the stationary solution $(0,\lambda)$.

Denote the eigenvalues of $L(\lambda)$ by $\mu_j(\lambda)$ $j=1,\ldots,n$. $\mu_j(\lambda)$ are continuous in λ and occur in conjugate pairs if they are not real. Then the eigenvalues of $I-\exp tL(\lambda)$ are $1-\exp t\mu_j(\lambda)$ and the determinant of $I-\exp tL(\lambda_o)$ is $\prod_1^n (1-\exp t\mu_j(\lambda_o))$. So, if $Re(\mu_j(\lambda_o)) \neq 0$, then for all $t > 0$, the corresponding term $1-\exp t\mu_j(\lambda_o)$ is not zero, while if for some j: $\mu_j(\lambda_o) = i\beta$ ($\beta \neq 0$ since $L(\lambda_o)$ is not singular) then for $t\beta \equiv 0\ [2\pi]$ one has $\det(I-\exp tL(\lambda_o)) = 0$.

Suppose $L(\lambda_o)$ has eigenvalues $\pm i\beta$ ($\beta > 0$) and let $1 \leq k_1 \leq \ldots \leq k_d$ be integers such that $ik_1\beta,\ldots,ik_d\beta$ are eigenvalues of $L(\lambda_o)$ (counted with multiplicity i.e. as roots of the characteristic polynomial). Let $t_o = 2\pi\beta^{-1}$. Then $I-\exp t_oL(\lambda_o)$ has a generalized kernel of dimension $2d$ and finding zeros of $F(x,\lambda,t)$ near $(0,\lambda_o,t_o)$ is a problem of bifurcation with two parameters λ and t. It is easy to see that to each eigenvalue $ik_j\beta$ of $L(\lambda_o)$ there corresponds an eigenvalue $\alpha_{k_j}(\lambda) + i\beta_{k_j}(\lambda)$ of $L(\lambda)$ with $\alpha_{k_j}(\lambda)$, $\beta_{k_j}(\lambda)$ real and continuous in λ with $\alpha_{k_j}(\lambda_o) = 0$, $\beta_{k_j}(\lambda_o) = k_j\beta$. Assume $\alpha_{k_j}(\lambda) \neq 0$ for $\lambda \neq \lambda_o$ $j = 1,\ldots,d$.

Definition 3.3: β (or $(0,\lambda_o,t_o)$) is said to have *odd algebraic multiplicity* if an odd number of the $\alpha_{k_j}(\lambda)$ change sign when λ passes through λ_o.

Theorem 3.4: Assume β has odd algebraic multiplicity then there is a positive number ρ_o so that for all ρ, with $0<\rho<\rho_o$, there is a positive number $r_o(\rho)$ such that for all r, $r < r_o(\rho)$, $F(x,\lambda,t) = 0$ has a solution in the ball B_o with $||x|| = r$ where $B_o = \{(x,\lambda,t) / ||x||^2 + |\lambda-\lambda_o|^2 + |t-t_o|^2<r^2+\rho^2\}$.

i.e. the corresponding periodic orbit is not stationary (the only stationary solutions near $(0,\lambda_o)$ are for $x = 0$).

Proof: Since $\alpha_{k_j}(\lambda) \neq 0$ for $\lambda \neq \lambda_o$, I-exp $tL(\lambda)$ is invertible for (λ,t) near (but not equal to) (λ_o,t_o) and has a determinant of constant sign (which may be supposed to be positive), so choose ρ_o such that for (λ,t) with $|\lambda-\lambda_o|^2+|t-t_o|^2=\rho^2 < \rho_o^2$ one has $||(I-\exp tL(\lambda))^{-1}|| < K(\rho)$ for some constant $K(\rho)$ depending on ρ. Choose $r_o(\rho)$ such that for $||x|| < r < r_o(\rho)$, $||g(x,\lambda,t)|| / ||x|| < 1/2K(\rho)$. Let then $H_r(x,\lambda,t) = (F(x,\lambda,t),||x||^2-r^2)$ be defined on B_o. From the choices of ρ and r, $H_r(x,\lambda,t)$ is not zero on S_o^{n+1}, the boundary of B_o, and can be deformed to $((I-\exp tL(\lambda))x,||x||^2-r^2)$. The proof will be completed (applying theorem 2.2 of chapter one) once the following has been established:

Lemma 3.5: $H_r(x,\lambda,t)$ as a map from S_o^{n+1} to $\mathbb{R}^{n+1}-\{0\}$ is not homotopically trivial if and only if β has odd algebraic multiplicity.

Proof: Let $S_o^1 = \{(\lambda,t) / |\lambda-\lambda_o|^2 + |t-t_o|^2 = \rho^2\}$. Denote by $GL(n)^+$ the set of $n \times n$ matrices with positive determinant and consider any continuous mapping $A(\lambda,t)$ from S_o^1 to $GL(n)^+$. Then $(A(\lambda,t)x,||x||^2 - r^2)$ is a map from S_o^{n+1} to $\mathbb{R}^{n+1}-\{0\}$ and if $A(\lambda,t,\tau)$ is a family of such mappings then the corresponding maps $(A(\lambda,t,\tau)x, ||x||^2 - r^2)$ are homotopic. So one has an induced homomorphism between $\pi_1(GL(n)^+)$ and $\pi_{n+1}(S^n)$. (It is easy to check that the group additions are compatible). Now it is well known that

$$A_o(\lambda,t) = \begin{bmatrix} \lambda-\lambda_o & t-t_o \\ -(t-t_o) & \lambda-\lambda_o \end{bmatrix}$$

generates $\pi_1(GL(2)^+) = Z$, and

$$\begin{bmatrix} A_o(\lambda,t) & 0 \\ 0 & I^{n-2} \end{bmatrix}$$

generates $\pi_1(GL(n)^+) = Z_2$ while $(A_o(\lambda,t)x, ||x||^2 - r^2)$ is homotopic to the Hopf map (Theorem 2.6 of chapter one), generator of $\pi_3(S^2) = Z$, and its suspension generates $\pi_{n+1}(S^n) = Z_2$. So the mapping defined above is an isomorphism. Then Alexander and Yorke have proved (Lemma 8.1) that the class of $I-\exp L(\lambda)$ in $\pi_1(GL(n)^+)$ is not zero if and only if β has odd algebraic multiplicity. For completeness, a proof of this lemma is given in the appendix.

Remark 3.6: One could consider the more general situation where $f(0,\lambda_o) = 0$ only at λ_o and $||f(x,\lambda) - L(\lambda)x|| \leq C(||x||^2 + |\lambda-\lambda_o|^{2p+1})$ for some p. Then if the corresponding estimates for $g(x,\lambda,t)$ and for $||(I-\exp tL(\lambda))^{-1}||$ allow the deformation from $(F(x,\lambda,t),||x||^2-r^2)$ to $((I-\exp tL(\lambda))x, ||x||^2 - r^2)$ on S_o^{n+1}, as in Section IV of chapter one, then the bifurcation result of theorem 3.4 is still valid, the trivial solution $(0,\lambda,t)$ being replaced by (x,λ,t) with $|\lambda-\lambda_o|^2 + |t-t_o|^2 = \rho^2$ where ρ will be of the form $(Mr)^{1/k}$ for some k.

Remark 3.7: If $\alpha_{k_j}(\lambda) = 0$ for λ near λ_o then $I - \exp tL(\lambda)$ will be singular on the curve $t_j(\lambda) = 2\pi\beta_{k_j}^{-1}(\lambda)$ with $t_j(\lambda_o) = t_o$. Since there are at most d such curves one could try to find a solution with t as a function of λ on a curve $t(\lambda)$ transversal to the above curves at (λ_o, t_o). Then on $S_o^n = \{(x,\lambda) / ||x||^2 + |\lambda-\lambda_o|^2 = r^2 + \rho^2\}$ with appropriate choices for r and ρ, $H_r(x,\lambda,t) = (F(x,\lambda,t) \; ||x||^2 - r^2)$ is not zero and one would have to compute its degree. But this degree is always zero since the determinant of $I - \exp t(\lambda)L(\lambda)$ does not change sign as λ passes through λ_o so one has to know more on the dependance of $g(x,\lambda,t(\lambda))$ on λ and try to apply the analogues of corollaries 3.8 and 3.9 of chapter one.

Remark 3.8: If $i\beta$ has even multiplicity but $L(\lambda_o)$ has a multiple of $i\beta$ as eigenvalue with odd multiplicity one will get bifurcation from that multiple. However it is possible that there are no periodic solutions with period near $2\pi/\beta$ as the following example, due to C.L. Siegel and reproduced by Alexander and Yorke in [0] (example 2.9), shows. Let:

$$\dot{x}_1 = \lambda x_1 + x_2 + x_1x_3 - x_2x_4$$

$$\dot{x}_2 = -x_1 + \lambda x_2 - x_2x_3 - x_1x_4$$

$$\dot{x}_3 = -2\lambda x_3 - 2x_4 + \tfrac{1}{2}(x_1^2 - x_2^2)$$

$$\dot{x}_4 = 2x_3 - 2\lambda x_4 - x_1x_2$$

If $p = x_1^2 + x_2^2$ and $q = x_3^2 + x_4^2$ then: $\dot{p} - 2\dot{q} = 2\lambda(p+4q)$. So, if $p+4q$ is positive, integration of the periodic functions over a period yields $\lambda = 0$ (The left hand side must integrate to 0). Moreover $\ddot{p} = p(p+4q)$ which implies, by the same reasoning, $p \equiv 0$.

So the stationary solutions are: $x_i = 0 \quad i = 1,\ldots,4$ and the only periodic solutions are: $x_1 = 0$, $x_2 = 0$, $x_3 = a\cos 2t - b\sin 2t$, $x_4 = a\sin 2t + b\cos 2t$ of period π. Here $L(\lambda)$ is the linear part of the system, with eigenvalues $\lambda \pm i$ and $-2\lambda \pm i$, so that the periodic solutions bifurcate from $(0,0,\pi)$ corresponding to $2i$. Note that for $\lambda = 0$ the system is Hamiltonian with

$$H = \tfrac{1}{2}(x_1^2 + x_2^2) - x_3^2 - x_4^2 + x_1x_2x_3 + \tfrac{1}{2}(x_1^2 - x_2^2)x_4.$$

<u>Example 3.9</u>: In case of even multiplicity there may be no bifurcation:

The system:

$$\begin{cases} \dot{x}_1 = \lambda^2 x_1 + x_2 + (x_1^2 + x_2^2)x_1 \\ \dot{x}_2 = -x_1 + \lambda^2 x_2 + (x_1^2 + x_2^2)x_2 \end{cases}$$

verifies: $(x_1^2 + x_2^2)' = 2(\lambda^2 + x_1^2 + x_2^2)(x_1^2 + x_2^2)$ so that the only periodic solution is: $x_1 = x_2 = 0$. Here $L(\lambda)$ has eigenvalues $\lambda^2 \pm i$ so that there is no crossing.

<u>Example 3.10</u>: If there are two crossings one may have no bifurcation.

Consider:

$$\dot{x}_1 = \lambda x_1 + x_2$$

$$\dot{x}_2 = -x_1 + \lambda x_2 + x^3(x_1^2 + x_3^2)$$

$$\dot{x}_3 = -\lambda x_3 + x_4$$

$$\dot{x}_4 = -x_3 - \lambda x_4 - x_1(x_1^2 + x_3^2)$$

Then $(x_2x_3 - x_1x_4)' = (x_1^2 + x_3^2)^2$ so that integration over a period will imply: $x_1 = x_3 \equiv 0$, and so $x_i = 0 \quad i = 1,\ldots,4$ is the only periodic solution. Here $L(\lambda)$ is the linear part with eigenvalues: $\pm\lambda\pm i$, so that i has multiplicity 2.

Remark 3.11: A global version of theorem 3.4 will be proved in section III of the next chapter.

CHAPTER THREE

SOLUTIONS IN THE LARGE

The two preceding chapters were concerned with a local theory of bifurcation. A natural question to ask is what happens to these branches when they leave the neighborhood of the bifurcation point: Do they run off to infinity, do they come back, do they stop somewhere and if so, where? Very little information was needed for the existence of a bifurcating branch and consequently very little is known about it, nevertheless one can give a partial answer to these questions. This chapter gives a generalization of the important and now well known theorem of P.H. Rabinowitz, [26] and [27], using again topological techniques.

Several problems arise in the study of the global situation. In fact, in (2): $Ax - T(\lambda)x = g(x,\lambda)$, where A is a Fredholm operator, the bahavior of the linear part $A - T(\lambda)$ is unknown for λ large; for example, is 0 an isolated point in the $T(\lambda)$ spectrum of A, is $A - T(\lambda)$ a Fredholm operator for all λ ? (This is not true in general and one has to consider the essential $T(\lambda)$ spectrum of A). Moreover the set of solutions of (2) may be very difficult to study, unless for example it is known that on any bounded domain this set is relatively compact. Finally one needs good topological tools valid for infinite dimensional spaces, such as generalized cohomologies for maps of the form Id - compact, k-set contractions or σ-proper Fredholm maps.

So here (2) will take the form:

$$(I - T(\lambda))x = g(x,\lambda) \tag{13}$$

where x belongs to E, a real or complex Banach space.
$T(\lambda)$ is a compact operator from $E \times \mathbb{K}$ to E depending analytically on λ and it is assumed that there exist λ_o in $\mathbb{K}$ such that $I - T(\lambda_o)$ is invertible.
$g(x,\lambda)$ is a compact (continuous) map from some domain D, in $E \times \mathbb{K}$, into E, $((0,\lambda_o)$ is assumed to be in $D)$. Furthermore:

$$||g(x,\lambda)|| = o(||x||) \quad \text{as} \quad ||x|| \to 0$$

for x in D, uniformly on compact λ sets.

Then for each λ, $T(\lambda)$ is compact and $I - T(\lambda)$ is a Fredholm operator of index zero (Goldberg [14], Theorem V.2.1, p. 117) and, using lemma 6.3 of chapter one, the set U of λ's in $\mathbb{K}$ for which $I - T(\lambda)$ is a Fredholm operator is connected and in fact is $\mathbb{K}$ itself. Moreover, since $I - T(\lambda_o)$ is invertible, $I - T(\lambda)$ is invertible for all λ except for a discrete set of points called the characteristic values of $I - T(\lambda)$.

<u>Remark 1</u>: In these conditions bifurcation occurs only at characteristic values. In fact, for any other λ, (13) is equivalent to $x = (I-T(\lambda))^{-1} g(x,\lambda)$ which should be solvable for x arbitrarily small. Since the right hand side is $o(||x||)$ this is not possible unless x is zero.

<u>Remark 2</u>: Suppose one is in the situation of the preceding chapters, i.e. with $A - T(\lambda)$ and 0 as an isolated $T(\lambda)$ - eigenvalue of A. Then there exists λ_1 small such that $A - T(\lambda_1)$ is invertible so that (2) becomes:
$(A-T(\lambda_1))x - (T(\lambda) - T(\lambda_1))x = g(x,\lambda)$ which is equivalent to:
$x - (A-T(\lambda_1))^{-1}(T(\lambda) - T(\lambda_1))x = (A-T(\lambda_1))^{-1}g(x,\lambda)$.
So if $(A-T(\lambda_1))^{-1}(T(\lambda) - T(\lambda_1))$ and $(A-T(\lambda_1))^{-1}g(x,\lambda)$ are

compact, one is in the situation of (13).

§ I : INFINITE DIMENSIONAL COHOMOLOGY

In the next section maps $I - F$, of the form $\{(I-T(\lambda))x - g(x,\lambda), ||x||^2 - r^2\}$, will be defined on some open set Ω in D such that, on the boundary $\partial\Omega$, $I - F$ is not zero and one will compute a topological invariant for $I - F$, in one case its Leray-Schauder degree, in the other, the complex case, $I - F$ maps $\partial\Omega$ into a subspace of $E \times \mathbb{C}$, so an extension of the Hopf map argument, stable cohomotopy, is necessary.

So, let E be a real Banach space (if E is complex, one forgets its complex structure and defines the analogous theories as in Cronin [7] and [8]). A fixed orientation is given to E by finite dimensional subspaces E_n of E such that E_n has dimension n, $E = E_n \oplus E^{\infty-n}$, $E_n \subset E_{n+1}$, $E_{n+1} = E_n \oplus \mathbb{R}$, $E^{\infty-n} = E^{\infty-n-1} \oplus \mathbb{R}$. Let

$$P^{\infty-n} = E^{\infty-n} - \overline{\mathbb{R}^+} = \{x \in E^{\infty-n}, x = x_1 \oplus r, x_1 \in E^{\infty-n-1}, r \in \mathbb{R}, r < 0\}$$

Furthermore one defines the Leray-Schauder category of E, $L{\cdot}S(E)$, whose objects are closed bounded subsets of E and whose morphisms are compact vector fields of the form I-compact, and homotopies $x - F(x,t)$ with F compact. Given A and X in $L{\cdot}S(E)$, A contained in X, one denotes the set of (compact) homotopy classes of maps $I - F$, F compact, with:

$$I - F : \begin{cases} X \to E^{\infty-n} - \{0\}, \\ A \to P^{\infty-n} \end{cases}$$

by $\Pi^{\infty-n}(X,A)$ and $\Pi^{\infty-n}(X) = \Pi^{\infty-n}(X,\phi)$.

Using approximation of compact maps by finite dimensional ones, Geba [13] and Granas [17] have proved that $\Pi^{\infty-n}(X,A)$ can be given the algebraic structure of an abelian group. In fact let $X_\alpha = X \cap E_\alpha$, $A_\alpha = A \cap E_\alpha$, for E_α in $\mathcal{K}(E)$, where $\mathcal{K}(E)$ is a directed set of finite dimensional subspaces E_α containing E_{n+1}. Then it can be shown, (as for the definition of the Leray-Schauder degree), that for each f in $\Pi^{\infty-n}(X,A)$ there exists an integer α, $\alpha = n + m + 1$, and a map g : $(X_\alpha, A_\alpha) \to (E^{\alpha-n}-\{0\}, P^{\alpha-n})$ such that f restricted to E_α and g are homotopic. And since $(E^{\alpha-n} - \{0\}, P^{\alpha-n})$ is homotopically equivalent to $(S^{\alpha-1-n}, pt)$, g is an element of $\pi^{\alpha-1-n}(X_\alpha, A_\alpha)$, the m-th cohomotopy group of (X_α, A_α).

<u>Remark 1.1</u>: $\pi^m(X_\alpha, A_\alpha)$ is an abelian group for $\dim X_\alpha \leq 2m-2$; here the topological dimension for a paracompact space is used. The group operation, for CW - complexes, is defined as follows: (Spanier [33]) Let $f,g: (X_\alpha, A_\alpha) \to (S^m, pt)$ be approximated by simplicial maps. Then consider:

$$(X_\alpha, A_\alpha) \overset{d}{\to} (X_\alpha, A_\alpha) \times (X_\alpha, A_\alpha) \overset{f \times g}{\to} (S^m, pt) \times (S^m, pt)$$

where d is the diagonal map. Then $(f \times g)d$ is a map from a $2m - 2$ complex to a complex of dimension $2m$. Standard approximation techniques enables one to deform this map so that its image lies in the $2m - 2$ skeleton of the range space and, in view of the particular form of $(S^m, pt) \times (S^m, pt)$, to $S^m V_{pt} S^m$ two copies of S^m wedged at the point pt. Then one follows this deformation by the map ω :

$S^m V_{pt} S^m \to (S^m, pt)$ defined as $\omega(x, pt) = \omega(pt, x) = x$.

The extra dimension is necessary in order to show that this

"addition" is commutative, and independent of the representative in the homotopy class of $f : (X_\alpha \times [0,1]$ has dimension at most $2m - 1)$. The inverse of an element is given by changing the orientation of S^m, i.e. by following f with a map of degree -1. Finally the trivial element is represented by $e : (X_\alpha, A_\alpha) \to (pt,pt)$.

A Mayer-Vietoris homomorphism gives a relation between $\pi^{\alpha-1-n}(X_\alpha, A_\alpha)$ and $\pi^{\alpha'-1-n}(X_{\alpha'}, A_{\alpha'})$, for $\alpha' \geq \alpha$ and $\alpha \geq 2n + 4$ (i.e. $m \geq n + 3$), so that one can identify $\Pi^{\infty-n}(X,A)$ with $\Sigma^{\infty-n}(X,A)$ the stable cohomotopy group of (X,A) direct limit (as α goes to ∞) of the groups $\pi^{\alpha-1-n}(X_\alpha, A_\alpha)$ and so the former set inherits all the algebraic properties of the later, in particular:

<u>Theorem 1.2: Geba-Granas.</u>

$\Pi^{\infty-n}$ <u>defines a generalized cohomology functor for</u> L·S(E), <u>that is</u>: If $f = I - F : (X,A) \to (Y,B)$, <u>then</u> f <u>induces a homomorphism</u> $f^* : \Pi^{\infty-n}(Y,B) \to \Pi^{\infty-n}(X,A)$ <u>by</u> $f^*[g] = [gof]$ <u>and</u>:

1) $(Id)^* = Id$.
2) $(fg)^* = g^*f^*$
3) <u>There exists a map</u> $\delta : \Pi^{\infty-n}(A) \to \Pi^{\infty-n+1}(X,A)$ <u>such that</u> $f^*\delta = \delta f_o^*$, <u>where</u> f_o <u>is the restriction of</u> f <u>to</u> A.
4) $\overset{\delta}{\to} \Pi^{\infty-n}(X,A) \to \Pi^{\infty-n}(X) \to \Pi^{\infty-n}(A) \overset{\delta}{\to} \Pi^{\infty-n+1}(X,A) \to \ldots \to \Pi^{\infty-o}(A)$ <u>is an exact sequence (where all arrows are either</u> δ <u>or induced by inclusions</u>).
5) <u>If</u> f <u>and</u> g <u>are compactly homotopic</u> $(x-F(x,t))$ <u>then</u> $f^* = g^*$.

6) <u>Strong excision</u>: <u>If</u> $X = A \cup B$, <u>with</u> X,A,B <u>in</u> $L^{\cdot}S(E)$ <u>then</u> $\Pi^{\infty-n}(X,B) \cong \Pi^{\infty-n}(A,A \cap B)$.

7) $\Pi^{\infty-n}$ (point) = 0.

<u>Remark 1.3</u>: The neutral element of $\Pi^{\infty-n}(X)$ may be characterized as the representative of all inessential compact vector fields from X to $E^{\infty-n} - \{0\}$ (Granas [17], p. B.VI.3). Recall that a compact vector field $f = I - F$, from X to $E^{\infty-n} - \{0\}$ is <u>inessential</u> if for all Y in $L^{\cdot}S(E)$, Y containing X, f extends to $\tilde{f} = I - \tilde{F} : Y \to E^{\infty-n} - \{0\}$. Then, using the homotopy extension theorem, any map which is compactly homotopic to an inessential map is itself inessential. Granas [16] Theorem 22.

<u>Remark 1.4</u>: One can recover the Leray-Schauder degree theory. In fact there is an Alexander-Pontrjagin duality between $\Pi^{\infty-n}(X)$ and $\Sigma_n(E-X)$ the stable homotopy group of $E - X$. In particular $\Pi^{\infty}(X) \cong \Sigma_o(E-X)$, so that $\Pi^{\infty}(X)$ is isomorphic to as many copies of Z as there are bounded components of $E - X$, and any compact vector field $f = I - F$ from X to $E - \{0\}$ has a class : $[f] = \Sigma m_i \alpha_i$, where α_i is represented by a mapping $x - x_i$, for x in X and x_i fixed in the i-th bounded component of $E - X$. Note that if m_i is not zero then any extension of f to the corresponding bounded component U_i of $E - X$ has a zero: If not f represents an element of $\Pi^{\infty}(X \cup U_i)$ which, due to the above duality, has one less component. So, as a consequence of the compactness of F and the boundedness of X, only a finite number of m_i are not zero: If not any compact extension of F to a ball containing X has an infinite sequence of fixed points each one in a different component of $E - X$ and from which one can extract a subsequence converging also to a fixed point which cannot belong to X since F is fixed point free on X; but

this fixed point cannot be in any U_i since these sets are disjoint and open, $E - X$ being locally connected.

So if X is the boundary of an open bounded set Ω in E and f is a compact vector field from Ω to E non-vanishing on X, one defines the degree of f with respect to Ω, at 0, as $\deg(f,\Omega,0) = \Sigma m_i$, where this sum is over the connected components of Ω. This degree has all the properties of the Leray-Schauder degree.

<u>Remark 1.5</u>: If S is a sphere in E, then $\Pi^{\infty-o}(S) = Z$, $\Pi^{\infty-1}(S) = Z_2$ generated by the stable Hopf map as can be seen from Example 2.4 in chapter one, $\Pi^{\infty-2}(S) = Z_2$, $\Pi^{\infty-3}(S) = Z_{24}\ldots$ Geba [13].

<u>Remark 1.6</u>: $\Pi^{\infty-n}(\overline{B}) = 0$ if B is a ball in E. In fact if $f(x) = x - F(x)$ maps $\overline{B}$ into $E^{\infty-n} - \{0\}$, f can be extended to a map from E to $E^{\infty-n} - \{0\}$ by defining $\tilde{f}$ outside $\overline{B}$ by: $\tilde{f}(x) = x - \frac{||x||}{R} F(\frac{x}{||x||} R)$ where R is the radius of B, assuming that B is centered at the origin.

<u>Remark 1.7</u>: An immediate application of this theory is to extend the results of chapter one to the following case: Suppose that in (2) : $(A-T(\lambda))x = g(x,\lambda)$, $g(x,\lambda)$ is not assumed to be Lipschitz continuous as in condition (1.6) of chapter one but that only: $||g(x,\lambda)|| \leq C(||x||^2 + |\lambda|^{2n+1})$ for some n, and suppose that after the reduction

$$\begin{cases} (10) : x_2 - (I-KQT(\lambda))^{-1}KQT(\lambda)x_1 - (I-KQT(\lambda))^{-1}KQg(x_1+x_2,\lambda) \\ (11) \quad (I-Q)T(\lambda)(I-KQT(\lambda))^{-1}x_1 + (I-Q)(I-T(\lambda)KQ)^{-1}g(x_1+x_2,\lambda), \end{cases}$$

$(I-KQT(\lambda))^{-1}KQg(x_1+x_2,\lambda)$ is a compact map from $B \times \mathbb{K}$ into B.

Assuming that 0 is an isolated $T(\lambda)$ eigenvalue of A, in the sense of section VI of chapter one, then the results of theorem 6.5 of that chapter hold: In fact, on the ball $\{||x_2|| < r, ||x_1||^2 + |\lambda|^2 < r^2 + (Mr)^{2/n}\}$, a solution of (10) gives $||x_2|| = o(r)$ and (10), (11) with the side equation $||x_1||^2 - r^2$ insure that there is no solution on the boundary of the ball. So one can deform $((10), (11), ||x_1||^2 - r^2)$ to $(x_2, B(\lambda)x_1, ||x_1||^2 - r^2)$ which is the suspension (or using the product theorem in the real case) of the map $(B(\lambda)x_1, ||x||^2 - r^2)$ on the sphere $\{||x_1||^2 + |\lambda|^2 = r^2 + (Mr)^{2/n}\}$ which was studied in extenso in chapter one.

§ II : GLOBAL SOLUTIONS

One is now ready for the extension of Rabinowitz' theorem [26], [27]. Let S be the closure in D of the non-trivial solutions of (13) (i.e. $x \neq 0$). And suppose that λ_o is a characteristic value of $I - T(\lambda)$ at which bifurcation takes place. Let $\mathcal{S}$ be the connected component of S containing $(0,\lambda_o)$. Note that the only trivial solutions in S are the points $(0,\lambda)$ where λ is a characteristic value of $I - T(\lambda)$. For the sake of clarity, three different cases will be studied: The first for $I - \lambda T$ and $\mathbb{K} = \mathbb{R}$, the second for $I - \lambda T$ and $\mathbb{K} = \mathbb{C}$, and the last for $I - T(\lambda)$.

A) $\mathbb{K} = \mathbb{R}$, $I - \lambda T$.

Theorem 2.1: $\mathcal{S}$ is either: i) Not compact in D (and if $D = E \times \mathbb{R}$, $\mathcal{S}$ is unbounded) or ii) $\mathcal{S}$ is bounded in D and then $\mathcal{S}$ connects a finite number of points $(0,\lambda_j)$, λ_j characteristic value, with the following property: In $\mathbb{R}$ order the

<u>characteristic values</u> λ_i, $i = 1,\ldots,p$ <u>of odd multiplicity, such that</u> $(0,\lambda_i)$ <u>belongs to</u> $\mathcal{S}$ (λ_o <u>may be one of these)</u>: $\lambda_1 < \lambda_2 < \ldots < \lambda_p$. <u>Let</u> n_1 <u>be the number of characteristic values</u> λ, <u>of odd multiplicity, such that</u>: $\lambda_1 < \lambda < \lambda_2$, n_2 <u>those between</u> λ_2 <u>and</u> $\lambda_3,\ldots,n_{p-1}$ <u>such that</u> $\lambda_{p-1} < \lambda < \lambda_p$ (n_j <u>may be</u> 0), <u>then</u>:

2.2:[(1)]
$$\begin{cases} 1 = (-1)^{n_1} - (-1)^{n_1+n_2} + \ldots + (-1)^{k+1}(-1)^{\sum_{i=1}^{k} n_i} \\ + \ldots + (-1)^p (-1)^{\sum_{i=1}^{p-1} n_i} \end{cases}$$

<u>Consequences 2.3</u>:

1) 2.2 implies that p is even.
2) If $p = 0$, λ_o is of even multiplicity. So if λ_o is of odd multiplicity $\mathcal{S}$ connects $(0,\lambda_o)$ to an odd number of points $(0,\lambda)$ with λ of odd multiplicity. Rabinowitz remarked this fact in [27] and proved in [26] that in this case $p \geq 2$.
3) $p = 2$, then n_1 is even
4) $p = 4$, the only possible combinations are: n_1 even, n_2 even, n_3 even; n_1 even, n_2 odd, n_3 even; n_1 odd, n_2 even, n_3 odd.
5) p even: all n_i even (in particular 0) is an admissible situation.

<u>Remark 2.4</u>: Here multiplicity means $\dim\left(\bigcup_{m=1}^{\infty} \ker(I-\lambda T)^m\right)$.

This will not be the case in the general situation of section C.

---

(1) E.N. Dancer (Indiana U. Math. J. Vol.23 No. 11 (1974)) has derived an analogeous formula, using Rabinowitz' presentation.

The proof of the theorem will follow from the computation of a local index. Precisely, let λ_o be a characteristic value of $I - \lambda T$, so λ_o is an isolated characteristic value. Then for ρ small enough, $\rho > 0$, let $i(\lambda_{o-})$ be the index of $(I-(\lambda_o-\rho)T)$ and $i(\lambda_{o+}) = \text{index } (I-(\lambda_o+\rho)T)$. Recall that $i(\lambda_{o-}) = (-1)^{\beta}$ where β is the sum of the multiplicities of characteristic values $\mu\lambda_o$ of $I - \mu\lambda_o T$, for $0 < \mu < 1$ if $\lambda_o > 0$ and for $0 < \mu \leq 1$ if $\lambda_o < 0$. Krasnosel'skii [22]. So $i(\lambda_{o+}) = (-1)^m \, i(\lambda_{o-})$, m multiplicity of λ_o.

<u>Lemma 2.5</u>: Consider $H_r(x,\lambda) = ((I-T(\lambda))x-g(x,\lambda),||x||^2-r^2)$ from a neighborhood of $(0,\lambda_o)$ in D into $E \times \mathbb{R}$. Then, for ρ and r small enough and $B = \{(x,\lambda) \,/\, ||x||^2 + |\lambda-\lambda_o|^2 < r^2+\rho^2\}$, degree $(H_r(x,\lambda),B,(0,0))$ is defined and equal to $i(\lambda_{o-})-i(\lambda_{o+})$.

<u>Proof</u>: Choose ρ so small that $I - (\lambda_o \pm t\rho)T$ is invertible for all $t \in]0,1]$ and choose r so small that the only solution x, $||x|| \leq r$, of $(I-(\lambda_o\pm\rho)T)x - g(x,\lambda_o\pm\rho) = 0$ is $x = 0$ (use Remark 1). So it is clear that on $\partial B = \{(x,\lambda) \,/\, ||x||^2 + |\lambda-\lambda_o|^2 = r^2 + \rho^2\}$, $H_r(x,\lambda) = 0$ has no solution. ($||x||^2 - r^2 = 0$ implies $\lambda = \lambda_o\pm\rho$, and the choice of r and ρ contradicts the fact that $x - (\lambda_o\pm\rho)Tx = g(x,\lambda)$). Similarly the deformation $((I-\lambda T)x - (1-t)g(x,\lambda)$, $(1-t)(||x||^2-r^2) + t(\rho^2-|\lambda-\lambda_o|^2)$ is valid since, on the boundary, the second equation is $||x||^2 - r^2$. So the degree of $(H_r(x,\lambda),B,(0,0)) = \text{degree } ((I-\lambda T)x, \rho^2-|\lambda-\lambda_o|^2, B, (0,0))$. Now the inverse image of $(0,0)$ under this map is $\lambda = \lambda_o\pm\rho$ and hence $x = 0$, so this degree is the sum of the two local indices at $(0,\lambda_o\pm\rho)$, and also the indices of the Frechet derivative of the map at these points (Krasnosel'skii [22], Theorem 4.8, Chapter II). But the Frechet derivative of

$((I-\lambda T)x, \rho^2 - |\lambda-\lambda_o|^2)$ at $(0,\lambda_o \pm \rho)$ is:

$$\begin{bmatrix} I - (\lambda_o \pm \rho)T & 0 \\ & \\ 0 & -2(\pm\rho) \end{bmatrix}$$

By the product theorem this index is: $i(\lambda_{o+}) \times (-1)$ for $\lambda_o + \rho$ and $i(\lambda_{o-}) \times (1)$ for $\lambda_o - \rho$. This proves the lemma.

Proof of the theorem: Since the intersection of the fixed point set of a compact map with a closed bounded set is compact, it is enough to consider the second part of the alternative in the theorem. So, supposing that $\mathcal{S}$ is compact, $\mathcal{S}$ connects $(0,\lambda_o)$ to a finite number of points $(0,\lambda_j)$ $j = 1,\ldots,N$ (using the fact that the set of characteristic values is discrete and Remark 1).

Let Ω be an open bounded subset of D containing $\mathcal{S}$ and such that (13) has no non-trivial solutions of its boundary $\partial\Omega$ and such that Ω contains no other points $(0,\lambda),\lambda$ characteristic value, than $(0,\lambda_j)$ $j = 1,\ldots,N$. The existence of such a set is proved in [27], Lemma 1.9.

So for any $r > 0$, degree $(H_r(x,\lambda),\Omega,(0,0))$ is defined and independent of r. Since Ω is bounded, for r large enough and for any x in Ω, $||x||^2 - r^2 < 0$, so that this degree is zero. For r very small, it has been seen in Remark 1 that $H_r(x,\lambda) = 0$ has a solution in Ω only near a characteristic value. Precisely choosing, as in Lemma 2.5, ρ such that $I - (\lambda_j \pm t\rho)T$ is invertible for t in $]0,1]$ and $j = 1,\ldots,N$, and such that $(0,\lambda_j \pm t\rho)$ is in Ω, the compactness of $S \cap \overline{\Omega}$ enables one to choose r small enough so that $H_r(x,\lambda) = 0$ can happen only inside small balls B_j in Ω of the form

$B_j = \{(x,\lambda) \; / \; ||x||^2 + |\lambda-\lambda_j|^2 < r^2 + \rho^2\}$. The addition property of the Leray-Schauder degree yields for such a choice of r:

$$\text{degree}(H_r(x,\lambda),\Omega,(0,0)) = 0 = \sum_{j=1}^{N} \text{degree}(H_r(x,\lambda),B_j,(0,0))$$

$$= \sum_{j=1}^{N} i(\lambda_{j-}) - i(\lambda_{j+}) \quad \text{by lemma 2.5.}$$

Since $i(\lambda_{j+}) = (-1)^{m_j} i(\lambda_{j-})$ only the characteristic values of odd multiplicity m_j count in this sum.

Let then $\lambda_1 < \ldots < \lambda_p$ be the characteristic values of odd multiplicity in this sum. The last formula becomes:

$0 = \sum_{i=1}^{p} i(\lambda_{i-})$. Now $i(\lambda_{i-}) = (-1)^{\beta_i}$, β_i is the sum of the multiplicities of $I - \mu\lambda_i T$ for $0 < \mu < 1$ if $\lambda_i > 0$ and $0 < \mu \leq 1$ if $\lambda_i < 0$. So the sum of the multiplicities of $I - \lambda T$ for λ in $[\lambda_1,\ldots,\lambda_i[$ is $\beta_i - \beta_1$ if $\lambda_1 > 0, \beta_1 - \beta_i$ if $\lambda_i < 0, \beta_1 + \beta_i$ if $\lambda_1 < 0$ and $\lambda_i > 0$.

In all case $(-1)^{\beta_i - \beta_1}$ corresponds to this sum. So that dividing the identity: $0 = \sum_{i=1}^{p} (-1)^{\beta_i}$ by $(-1)^{\beta_1}$, one gets:

$$1 + (-1)^{m_1+n_1} + (-1)^{m_1+n_1+m_2+n_2} + \ldots + (-1)^{\sum_{i=1}^{p-1} m_i+n_i} = 0 ,$$

where m_i are the multiplicities of λ_i, hence odd, and n_i were defined in the statement of the theorem. This gives 2.2.

B) $\mathbb{K} = \mathbb{C}$, $I - \lambda T$.

In the complex case $H_r(x,\lambda)$ is a mapping from D, in $E \times \mathbb{C}$, into $E \times \mathbb{R}$ a lower "dimensional" space, so here the

Geba-Granas theory will be used.

Theorem 2.6: $\mathcal{S}$ is either i) Not compact in D (if $D = E \times \mathbb{C}$, $\mathcal{S}$ is unbounded) or ii) $\mathcal{S}$ is bounded in D and then $\mathcal{S}$ connects a finite number of points $(0,\lambda_i)$, λ_i characteristic value, such that, if $\lambda_1,\ldots,\lambda_p$ are those of odd multiplicity, then p is even. (So that if λ_o is of odd multiplicity, $\mathcal{S}$ connects $(0,\lambda_o)$ to an odd number of points $(0,\lambda_i)$, λ_i of odd multiplicity.) (1)

(1) Dancer (personal communication) has pointed out to the author that in case $g(x,\lambda)$ is complex analytic then $\mathcal{S}$ is always unbounded: In fact if $\mathcal{S}$ is bounded, constructing Ω as before and letting $\Omega_\lambda = \{x \mid (x,\lambda) \in \Omega\}$, then for λ large enough: $\deg((13),\Omega_\lambda,0)=0$. By restricting λ to belong to a straight line passing through λ_o and containing no other characteristic value in $\mathcal{S}$ then, for λ close (but not equal) to λ_o, the above degree is equal to $\deg((13),\Omega_\lambda - B_\lambda,0)$ where B_λ is a small ball in the x-space isolating the zero solution (this equality comes from the construction of Ω: see [26]). But then (13) has no solution in $\Omega_\lambda - B_\lambda$: J.T. Schwartz, "Compact analytical mappings of B-spaces and a theorem of Jane Cronin", Comm. Pure Appl. Math. 16 (1963), 253-260. This means that on this line any solution in $\mathcal{S}$ must have $\lambda = \lambda_o$. But then $\deg((13),\Omega_{\lambda_o},0) = \mathrm{Index}(I-\lambda T)=1$ by homotopy invariance. From the properties of the degree for complex analytic maps given in the above reference, 0 is the only solution in Ω_{λ_o} and $I-\lambda_o T$ is invertible which contradicts the fact that λ_o is a characteristic value.

The proof of the theorem will again follow from the computation of a local invariant.

Lemma 2.7: There exists ρ and r positive and small enough such that on $S = \{(x,\lambda) \,/\, ||x||^2 + |\lambda-\lambda|^2 = r^2 + \rho^2\}$ the stable homotopy class of $H_r(x,\lambda) = \{(I-\lambda T)x - g(x,\lambda),\ ||x||^2 - r^2\}$ is defined and equal to $\Sigma(m\eta)$, m multiplicity of λ_o, η the Hopf map, Σ stable suspension.

Proof: As for lemma 2.5, choose ρ so small that $(I-(\lambda_o+\rho e^{i\theta})T)^{-1}$ is bounded by a constant M, for all θ in $[0,2\pi]$. (M is given by a continuity argument.) Then choose r so small that the only solution of $(I-(\lambda_o+\rho e^{i\theta})T)x - g(x,\lambda_o+\rho e^{i\theta}) = 0$ is $x = 0$, if $||x|| \leq r$, (again possible by Remark 1 for r such that $M \times o(r)/r < 1$). Moreover on the above sphere $H_r(x,\lambda)$ can be deformed to $((I-\lambda T)x,\ \rho^2-|\lambda-\lambda_o|^2)$, (as in the real case). Let $f(t) : [0,1] \to \mathbb{C}$ be a path such that $f(0) = 0$, $f(1) = 1$ and $f(t)\lambda_o$ is a path from 0 to λ_o which avoids all characteristic points. Let δ be the distance of this path to the set of characteristic values (excluding λ_o). Choose $\rho < \delta/\max|f(t)|$ and write E as $R(I-\lambda_oT)^\alpha \oplus \ker(I-\lambda_oT)^\alpha$ as in section III of chapter one, α being the ascent of $I - \lambda_oT$, m the dimension of the generalized kernel. So $x = x_1 \oplus x_2$ with x_2 in $R(I-\lambda_oT)^\alpha$; and the deformation of $H_r(x,\lambda)$ can be written as:
$((I-\lambda T)x_2 \oplus (I-\lambda T)x_1\ ,\ \rho^2-|\lambda-\lambda_o|^2)$. On S deform this map via $((I-f(t)\lambda T)x_2 \oplus (I-\lambda T)x_1,\ \rho^2-|\lambda-\lambda_o|^2)$. This deformation, of the form I-compact, is possible since, if for (x,λ) in S the map is null, then by the choice of ρ, $x_1 = 0$ and x_2 belongs to $\ker(I-f(t)\lambda T)$. But the choice of δ and ρ such that $|f(t)|\ \rho < \delta$ implies that this is not possible unless $f(t)\lambda = \lambda_o$, in which case x_2 is in $\ker(I-\lambda_oT)^\alpha \cap R(I-\lambda_oT)^\alpha = \{0\}$.
So $||x||^2$ cannot be r^2. Note that the side condition $||x||^2-r^2$ can now be deformed to $||x_1||^2-r^2$.
So the homotopy class of $H_r(x,\lambda)$ on S is the homotopy class

of $(x_2 \oplus (I-\lambda T)x_1, ||x_1||^2-r^2)$ i.e. the suspension of the class of $((I-\lambda T)x_1, ||x_1||^2-r^2)$ on $\{(x_1,\lambda) / ||x_1||^2 + |\lambda-\lambda_o|^2=r^2+\rho^2\}$. Finally, if one chooses a basis for $\ker(I-\lambda_o T)^\alpha$ such that $I - \lambda_o T$ is in Jordan form then, $I - \lambda T$ on a typical block is of the form: $1 - \lambda/\lambda_o$ on the diagonal, λ/λ_o in the upper diagonal. As in section IV of chapter one (proof of theorem 4.4) the off-diagonal terms can be deformed to zero and one is left with a matrix of the form given in (2.5) of chapter one with determinant $(1 - \frac{\lambda}{\lambda_o})^m$, and the condition $||x_1||^2 - r^2$. Theorem 2.6 of chapter I permits us to conclude the proof of this lemma.

Proof of the Theorem: Construct, as in the real case, the open bounded set Ω such that $(I-T(\lambda))x - g(x,\lambda)$ has no zero on its boundary, except $x = 0$. So for all $r > 0$, $H_r(x,\lambda)$ represents the same element in $\Pi^{\infty-1}(\partial\Omega)$. The orientation in $E \times \mathbb{C}$ is chosen such that $E^{\infty-2} = E$, $E^{\infty-1} = E \times \mathbb{R}$ ($\mathbb{R}$ real axis of $\mathbb{C}$). For r large enough, $||x||^2 - r^2 < 0$ in Ω, so in fact $H_r(x,\lambda)$ represents also an element in $\Pi^{\infty-1}(\overline{\Omega},\partial\Omega)$ $(P^{\infty-1} = E^{\infty-1}-\overline{\mathbb{R}}^+)$ which, by restriction i^*, gives an element (also represented by $H_r(x,\lambda)$ for r large) in $\Pi^{\infty-1}(\overline{\Omega})$ and finally, by restriction j^*, an element in $\Pi^{\infty-1}(\partial\Omega)$ (again $H_r(x,\lambda)$ for any $r > 0$ this time). But by theorem 1.2:

$$\Pi^{\infty-1}(\overline{\Omega},\partial\Omega) \xrightarrow{i^*} \Pi^{\infty-1}(\overline{\Omega}) \xrightarrow{j^*} \Pi^{\infty-1}(\partial\Omega)$$

is an exact sequence, in particular $j^*i^* = 0$. This implies that $H_r(x,\lambda) = 0$ or else, by Remark 1.3, $H_r(x,\lambda)$ is inessential with respect to $\partial\Omega$, for all r.

Choose r so small that any solution of $H_r(x,\lambda) = 0$ in Ω must lie inside balls $B_j = \{(x,\lambda) / ||x||^2 + |\lambda-\lambda_j|^2 < r^2 + \rho^2\}$, where ρ is taken as in the proof of theorem 2.1 and such that the conclusion of lemma 2.7 is valid for all $(0,\lambda_j)$ connected to $(0,\lambda_o)$ by $\mathcal{S}$.

Then, since $H_r(x,\lambda)$ is inessential with respect to $\partial\Omega$, let B be a ball in $E \times \mathbb{C}$, containing Ω, (so that $\overline{B}$ belongs to $L\cdot S(E \times \mathbb{C})$) and extend $H_r(x,\lambda)$ to $\overline{B}$. Define $\tilde{H}_r(x,\lambda)$ to be this extension on $\overline{B} - \Omega$ and $H_r(x,\lambda)$ on $\overline{\Omega}$. From this construction and from Remark 1.6, $\tilde{H}_r(x,\lambda)$ is still inessential with respect to S the boundary of B and $\tilde{H}_r(x,\lambda)$ maps

$\overline{B} - \bigcup_{j=1}^{N} B_j$ into $E^{\infty-1} - \{0\}$.

The last part of the proof consists in relating the global class of $\tilde{H}_r(x,\lambda)$ in $\Pi^{\infty-1}(S)$ (which is 0) to the local classes in $\Pi^{\infty-1}(S_j)$.

First $\tilde{H}_r(x,\lambda)$ restricted to $\bigcup_{j=1}^{N} S_j$ defines an element in

$\Pi^{\infty-1}(\bigsqcup_{j=1}^{N} S_j) \cong \bigoplus_{j=1}^{N} \Pi^{\infty-1}(S_j) \cong \bigoplus_{j=1}^{N} Z_2$ (the first isomorphism is clear since the B_j's are disjoint by construction and then one uses the first definition of $\Pi^{\infty-1}(X)$, the second was given in Remark 1.5). By lemma 2.7, $\tilde{H}_r(x,\lambda)|_{S_j}$ is represented by $\Sigma(m_j\ \eta)$, m_j the multiplicity of λ_j. Secondly it was noted that $\tilde{H}_r(x,\lambda)$ is not zero in $\overline{B} - \bigcup_{j=1}^{N} B_j$, so in the exact sequence:

$$(2.8) \qquad \Pi^{\infty-1}(\overline{B} - \bigsqcup_{j=1}^{N} B_j, \bigsqcup_{j=1}^{N} S_j) \xrightarrow{k^*} \Pi^{\infty-1}(\overline{B} - \bigsqcup_{j=1}^{N} B_j)$$

$$\overset{i*}{\to} \Pi^{\infty-1}(\bigcup_{j=1}^{N} S_j) \overset{\delta}{\to} \Pi^{\infty-0}(\overline{B} - \bigcup_{j=1}^{N} B_j, \bigcup_{j=1}^{N} S_j)$$

One has $[\tilde{H}_r(x,\lambda) \big| \bigcup_{j=1}^{N} S_j] = i*[H_r(x,\lambda) \big| \overline{B} - \bigcup_{j=1}^{N} B_j]$.

Now $i*$ is an isomorphism: In fact $\Pi^{\infty-n}(\overline{B} - \bigcup_{j=1}^{N} B_j, \bigcup_{j=1}^{N} S_j)$ is isomorphic to $\Pi^{\infty-n}(\overline{B}, \bigcup_{j=1}^{N} \overline{B}_j)$ by excision of $\bigcup_{j=1}^{N} B_j$ (in the notation of theorem 1.2: $X = \overline{B}$, $B = \bigcup_{j=1}^{N} \overline{B}_j$, $A = \overline{B} - \bigcup_{j=1}^{N} B_j$ (so that $A \cap B = \bigcup_{j=1}^{N} S_j$))

This last group is an element of the exact sequence:

$$\Pi^{\infty-n-1}(\bigcup_{j=1}^{N} \overline{B}_j) \overset{\delta}{\to} \Pi^{\infty-n}(\overline{B}, \bigcup_{j=1}^{N} B_j) \to \Pi^{\infty-n}(\overline{B}) .$$

As noted in Remark 1.6 and using the disjointness of the B_j's, both extreme groups are trivial so that, by exactness, the middle group is also zero. Hence exactness in (2.8) enables one to conclude that $i*$ is an isomorphism and that, if α_j, represented by $\Sigma\eta$ on S_j, generates $\Pi^{\infty-1}(S_j)$ then $\oplus\alpha_j$ generate $\Pi^{\infty-1}(\overline{B} - \bigcup_{j=1}^{N} B_j)$, and so $[\tilde{H}_r(x,\lambda) \big| \overline{B} - \bigcup_{j=1}^{N} B_j] = \oplus[\Sigma(m_j \eta)] = \oplus\, m_j\alpha_j$.

Finally, consider $\Pi^{\infty-1}(\overline{B} - \bigcup_{j=1}^{N} B_j) \overset{k*}{\to} \Pi^{\infty-1}(S)$ where $k*$ is induced by inclusion. Decompose E as $E = Y_2 \oplus \mathbb{C}x_o$, for some fixed element x_o in E, and write any x as $x = y_2 \oplus z x_o$.

Consider the map $(y_2, (\lambda-\lambda_j)z, ||x||^2 - r^2)$ on S_j. Its stable class is obviously the suspension of the Hopf map and so can be regarded as representing α_j. Now the same map is also non-trivial in $\Pi^{\infty-1}(S)$ while it is trivial in $\Pi^{\infty-1}(S_i)$ for i different from j (it is inessential with respect to $S_i : \lambda - \lambda_j \neq 0$ on $\bar{B}_i$). Hence, since k^* is induced by inclusion, $k^*(\alpha_j) = \alpha$ the generator of $\Pi^{\infty-1}(S)$ and k^* being a homomorphism $k^*(m_j\alpha_j) = m_j\alpha$, for all j, so

$$k^*(\oplus m_j\alpha_j) = \sum_1^N m_j\alpha = \left[\Sigma(\sum_1^N m_j\eta)\right] = 0, \text{ since } k^*(\oplus m_j a_j) =$$

$$k^*\left[\tilde{H}_r(x,\lambda)\Big|_{\bar{B} - \bigcup_1^N B_j}\right] = \left[\tilde{H}_r(x,\lambda)\Big|_S\right] \text{ which is inessential. So}$$

because $\Pi^{\infty-1}(S) \cong Z_2$, the last equation implies that $\sum_{j=1}^N m_j$ is even and so one has an even number of odd multiplicities m_j, thus proving the theorem. (1)

Remark 2.9: It is clear that the above argument can be reproduced for the real case, since it substitutes the addition property of the Leray-Schauder degree theory. (One has to be careful about orientations when relating the classes on S_j and S).

C) $\mathbb{K} = \mathbb{R}$ or $\mathbb{C}$, $I - T(\lambda)$.

Note here that the proofs of theorems 2.1 and 2.6 remain valid for this case that is, if $\mathcal{S}$ is bounded, then the "sum" of the

(1) K. Geba (preprint) has proved for $\Pi^{\infty-1}(\partial\Omega)$ an addition formula similar to the one for the Leray-Schauder degree. Thus the above computation may be avoided but one has to introduce framed bordism theory in Banach spaces.

local classes is equal to zero. So the difficulty is to compute these classes.

Recall that in section VI of chapter one a local class was also computed: That is, if λ_o is a characteristic value of $I - T(\lambda)$, the equation $(I-T(\lambda))x - g(x,\lambda)$ can be written as $Ax - C(\lambda)x - g(x,\lambda)$, where $A = I - T(\lambda_o)$ is a Fredholm operator of index zero and non-trivial kernel, $C(\lambda) = T(\lambda) - T(\lambda_o)$ is a compact operator, analytic in λ and vanishing at λ_o. Then it was shown that, if K is the pseudo inverse of A, the above equation is equivalent to:

$$\begin{cases} (10)\quad x_2 - (I-KQC(\lambda))^{-1}KQC(\lambda)x_1 - (I-KQC(\lambda))^{-1}KQg(x_1+x_2,\lambda) = 0 \\ (11)\quad (I-Q)C(\lambda)(I-KQC(\lambda))^{-1}x_1 + (I-Q)(I-C(\lambda)KQ)^{-1}\ g(x_1+x_2,\lambda) = 0 \end{cases}$$

where x is decomposed as $x = x_1 + x_2$, x_1 in ker A, x_2 in X_2 and Q is the projection on $R(A)$. As it was pointed out in Remark 1.7, λ_o is an isolated characteristic value and, if $B(\lambda)$ denotes $(I-Q)C(\lambda)(I-KQC(\lambda))^{-1}\Big|_{\ker A}$, $B(\lambda)$ has a determinant with dominant term $(\lambda-\lambda_o)^m$, where m was called the algebraic multiplicity of λ_o. Although the zero sets of (10), (11) and of the original equation are the same, one has to show that the local classes of the two different equations are the same.

<u>Theorem 2.10</u>: $\mathcal{S}$ <u>is either</u> i) <u>non compact in</u> D <u>(or if</u> $D = E \times \mathbb{K}$ $\mathcal{S}$ <u>is unbounded)</u>, <u>or</u> ii) $\mathcal{S}$ <u>is bounded and the conclusions of theorems 2.1 and 2.6 remain valid, i.e.</u>: If $\mathbb{K} = \mathbb{R}$: $\mathcal{S}$ <u>connects an even number of characteristic values of odd algebraic multiplicity and in fact for</u> $\lambda_1 < \ldots < \lambda_p$ <u>with</u> n_1 <u>being the number of</u> λ's <u>of odd algebraic multiplicity between</u> λ_1 <u>and</u> λ_2 <u>and so on ... one has</u>:

(2.2) $$1 = (-1)^{n_1} - (-1)^{n_1+n_2} + \ldots + (-1)^p (-1)^{\sum_1^{p-1} n_i}$$

If $\mathbb{K} = \mathbb{C}$: $\mathcal{S}$ connects an even number of characteristic values of odd algebraic multiplicity.

Remark 2.11: Note that the algebraic multiplicity at λ_o is different from the multiplicity of $I - T(\lambda_o)$. For example if $E = \mathbb{R}^2$, $T(\lambda) = \lambda T_1 + \lambda^2 T_2$, $T_1 = \begin{pmatrix} 1 & 0 \\ 0 & 2 \end{pmatrix}$, $T_2 = \begin{pmatrix} 0 & 0 \\ 0 & -1 \end{pmatrix}$

then $I - T(\lambda) = \begin{pmatrix} 1-\lambda & 0 \\ 0 & (1-\lambda)^2 \end{pmatrix}$ and $\lambda = 1$ has algebraic multiplicity 3 while $I - T(1) \equiv 0$ has multiplicity 2. The index of $I - T(1-\varepsilon)$ is 1, the index of $I - T(1+\varepsilon)$ is -1 and the degree of $(I-T(\lambda), ||x||^2 - r^2)$ near $(0,1)$ is 2.

Proof of the theorem: One needs only to compute the local class of $H_r(x,\lambda) = ((I-T(\lambda))x - g(x,\lambda), ||x||^2 - r^2)$ on a sphere $S_o = \{(x,\lambda) / ||x||^2 + |\lambda-\lambda_o|^2 = r^2 + \rho^2\}$ where r and ρ are chosen so small that λ_o is isolated and that $H_r(x,\lambda)$ can be deformed to its linear part $(I-T(\lambda), ||x||^2 - r^2)$. (As in the proof of Lemma 2.5, the second equation can be replaced by $\rho^2 - |\lambda-\lambda_o|^2$). Let Q be the projection of E onto $R(A) = R(I-T(\lambda_o))$ and writing x as $x_1 + x_2$, x_1 in ker A, one can express $((A-C(\lambda))x, \rho^2 - |\lambda-\lambda_o|^2)$ as:

(14) $$(Ax_2 - QC(\lambda)(x_1+x_2) \oplus (I-Q)C(\lambda)(x_1+x_2), \rho^2 - |\lambda-\lambda_o|^2).$$

Noting that (14) is of the form Id-compact, perform the following deformation on S_o:

$$((A-QC(\lambda))(x_1+x_2) \oplus (I-Q)C(\lambda)(x_1+(1-t)x_2 + tKQC(\lambda)(x_1+x_2)),$$
$$\rho^2 - |\lambda-\lambda_o|^2) .$$

This is a permissible deformation since if it is zero, then $tx_2 = tKQC(\lambda)(x_1+x_2)$ and the second part reads $(I-Q)C(\lambda)(x_1+x_2) = 0$, so that one is back to the original equation $((A-C(\lambda))x, \rho^2 - |\lambda-\lambda_o|^2)$ which has no zero in S_o. So the homotopy class of (14) is the class of:

$$(Ax_2-QC(\lambda)(x_1+x_2) \oplus (I-Q)C(\lambda)(x_1+KQC(\lambda)(x_1+x_2)), \rho^2-|\lambda-\lambda_o|^2)$$

Repeating this process n times one has to consider the class of:

$$\Big[Ax_2-QC(\lambda)(x_1+x_2) \oplus (I-Q)C(\lambda)\{(I+KQC(\lambda) + \dots + (KQC(\lambda))^n)x_1 + (KQC(\lambda))^n x_2\}, \rho^2 - |\lambda-\lambda_o|^2\Big]$$

where one recognizes the first n terms of $(I-Q)C(\lambda)(I-KQC(\lambda))^{-1}\Big|_{\ker A} = B(\lambda)$ where $\det B(\lambda) = (\lambda-\lambda_o)^m a(\lambda-\lambda_o)$, $a(0) \neq 0$. So for $n \geq m$ it is clear that, (as in the proof of theorem 4.4 of chapter one),

$$||(I-Q)C(\lambda)(I + \dots + (KQC(\lambda))^n)x_1|| \geq A|\lambda-\lambda_o|^{m-d+1} \; ||x_1||$$
(d dimension of $\ker A$),

while if $Ax_2 = QC(\lambda)(x_1 + x_2)$ then $||x_2|| \leq C|\lambda-\lambda_o| \; ||x_1||$ and $||(I-Q)C(\lambda)(KQC(\lambda))^n x_2|| \leq B|\lambda-\lambda_o|^{n+2} \; ||x_1||$, so that by the choice of n, for ρ small enough, one can deform the last map to $(Ax_2 \oplus B(\lambda)x_1, \rho^2 - |\lambda-\lambda_o|^2)$ on S_o.

Now consider the isomorphism from $E \times \mathbb{K}$ onto itself given by: $(y_2 = (I-T(\lambda_o))x_2, y_1 = Lx_1, \lambda = \lambda)$ where L is an isomorphism from $\ker A$ onto $\operatorname{coker} A$. Under this isomorphism (of the form I-compact), S_o is transformed into the sphere

$S_o' = \{(y_2 \oplus y_1, \lambda) \ / \ ||Ky_2 \oplus L^{-1} y_1||^2 + |\lambda-\lambda_o|^2 = r^2 + \rho^2\}$

and $((I-T(\lambda_o))x_2 \oplus B(\lambda)x_1, \rho^2 - |\lambda-\lambda_o|^2)$ into $(y_2 \oplus B(\lambda)L^{-1} y_1, \rho^2 - |\lambda-\lambda_o|^2)$. By contrafunctoriality of the cohomology theory, this isomorphism induces an isomorphism of groups from $\Pi^{\infty-n}(S_o')$ onto $\Pi^{\infty-n}(S_o)$. And since the last map has a non-trivial class if and only if the algebraic multiplicity of λ_o is odd, so does the former map and this settles the complex case.

Now in the real case, the reduction of (14) near λ_o to its Frechet derivative given in Lemma 2.5 is still valid, so that:

$$\text{degree}((14), B_o, (0,0)) = i(\lambda_o-\rho) - i(\lambda_o+\rho)$$

where $i(\lambda_o-\rho) = (-1)^{\beta-}$ β_- sum of multiplicities of $I - \mu T(\lambda_o-\rho)$ with $0 < \mu < 1$.

The above computation shows that $i(\lambda_o+\rho) = (-1)^m i(\lambda_o-\rho)$ since this degree is ± 2 if m is odd, 0 if m is even. Moreover suppose that λ is not a characteristic value, then $I - T(\lambda+\varepsilon) = I - T(\lambda) + \varepsilon D(\lambda,\varepsilon)$ where $D(\lambda,\varepsilon)$ is bounded. So for ε small enough $||I-T(\lambda+\varepsilon)|| \geq \frac{1}{2}||I-T(\lambda)||$ and so both have the same index with respect to any ball in the E space. This means that $i(\lambda)$ is constant except at characteristic values where it has a flip of sign only if the algebraic multiplicity is odd. Then starting from the first characteristic value of odd multiplicity in $\mathcal{S}$, the counting argument of theorem 2.1 finishes the proof of theorem 2.10.

Note

One can further deform $((I-T(\lambda_o))x_2 \oplus B(\lambda)x_1, \rho^2 - |\lambda-\lambda_o|^2)$ by decomposing E as $\ker(I-T(\lambda_o))^\alpha \oplus R(I-T(\lambda_o))^\alpha$ and by appropriate choice of P and Q (the parity of the multiplicity does not depend on P and Q: Theorem 6.4 in chapter one), the equation can be written as:

$$((I-T(\lambda_o))z_2 \oplus (I-T(\lambda_o))z_1 + B(\lambda)x_1,\ \rho^2 - |\lambda-\lambda_o|^2)$$

where x_2 was written as $x_2 = z_2 \oplus z_1$, z_2 in $R(I-T(\lambda_o))^\alpha$ on which $I - T(\lambda_o)$ is invariant.

So in the real case $(I-T(\lambda_o))z_2$ will give a contribution of $(-1)^\beta$, β sum of the multiplicities of $I - \mu T(\lambda_o)$ for $0 < \mu < 1$, while in the complex case it can be deformed to z_2 as in the proof of lemma 2.7. The finite dimensional part will give a contribution corresponding to the first exponent of its determinant (as in the proof of theorem 4.4 of chapter one).

If $I - T(\lambda_o)$ is in Jordan form on $\ker(I-T(\lambda_o))^\alpha$, this part will be of the form:

$$\begin{bmatrix} \begin{bmatrix} 0 & 1 & & \\ & \ddots & \ddots & \\ & & \ddots & 1 \\ b_{11} & \cdots & & 0 \end{bmatrix} & b_{12}\cdots\cdots & b_{13}\ \cdots \\ \vdots & \begin{bmatrix} 0 & 1 & & \\ & \ddots & \ddots & \\ & & \ddots & 1 \\ b_{21} & 0\cdots\ b_{22} & & 0 \end{bmatrix} & \vdots \\ & & b_{23}\ \cdots \\ \vdots & \vdots & \vdots \end{bmatrix}$$

which has a determinant $\pm\det B(\lambda)$.

Remark 2.12: If one starts with a Fredholm operator of positive index $i(A)$, with a compact pseudo inverse as it is the case for elliptic operators, by adding $i(A)$ equations (as in theorem 4.4 in chapter one) it is possible to get similar results. Also for several parameters, if the local indices are defined in some subspace of the parameter space, the application of the proof of theorem 2.6 (this time with $\Pi^{\infty-n}$) will give information on the global solutions.

§ III : THE HOPF BIFURCATION THEOREM: GLOBAL VERSION.

In this section the theorem of Alexander and Yorke [0] for global bifurcation of periodic orbits for autonomous systems will be proved. Note that in their article Alexander and Yorke derive other interesting results such as the variation of the period on the bifurcating branch and the Liapounov center theorem for Hamiltonian systems.

Recall that the problem of finding periodic orbits to the system $\dot{x} = f(x,\lambda)$ is equivalent to the existence of solutions of the equation $F(x,\lambda,t) = 0$, where F is a map from W an open set in $\mathbb{R}^{n+2}$ to M an open set in $\mathbb{R}^n$, the stationary points correspond to couples (x_o,λ_o) so that $F(x_o,\lambda_o,t) = 0$ for all t in $\mathbb{R}^+$.

Let P be the set $\{(x,\lambda,t) \ / \ F(x,\lambda,t) = 0$ and $f(x,\lambda) \neq 0\}$ and S the set $\{(x,\lambda,t) \ / \ f(x,\lambda) = 0\}$ of stationary points. It is easily seen that the points in $\overline{P} - P$ are in S. (Alexander-Yorke Lemma 11.1). See Note in the Appendix.

It is assumed that $(0,\lambda,t)$ belongs to S for all λ in Λ_o an open interval around λ_o and for all t in $\mathbb{R}^+$ and that $f_x(0,\lambda_o) = L(\lambda_o)$ has $i\beta$ as eigenvalue of odd algebraic multiplicity in the sense of definition 3.3 of chapter II. Under these conditions it has been seen in theorem 3.4 of chapter two that $H_r(x,\lambda,t) = (F(x,\lambda,t), ||x||^2 - r^2)$ as a map from S_o^{n+1} to $\mathbb{R}^{n+1}-\{0\}$ is not homotopically trivial, where $S_o^{n+1} = \{(x,\lambda,t) \;/\; ||x||^2 + |\lambda-\lambda_o|^2 + |t-t_o|^2 = r^2 + \rho^2\}$ and $t_o = 2\pi\beta^{-1}$.

Denote by $\mathcal{S}$ the connected component of $P\cup\{0,\lambda_o,t_o\}$ containing $(0,\lambda_o,t_o)$.

<u>Theorem 3.1</u>: $\mathcal{S}$ <u>is either</u> 1) <u>Not contained in any compact subset of</u> W <u>(and if</u> $W = \mathbb{R}^{n+1}\times\mathbb{R}^+$, $\mathcal{S}$ <u>is unbounded)</u>

or: 2) $\overline{\mathcal{S}}$ <u>contains a stationary point</u> $(x_1,\lambda_1,t_1) \neq (0,\lambda_o,t_o)$. <u>Note that, from the bifurcation result of chapter two, if</u> $x_1 = 0$, <u>then</u> $(\lambda_1 t_1)$ <u>cannot be close to</u> (λ_o,t_o).

<u>Proof</u>: This result due to Alexander and Yorke is now an easy consequence of the stable cohomotopy theory used in this chapter.

Suppose that none of the two branches of the above alternative is true. Then $\mathcal{S}$ is closed in $\overline{P}$, since $\overline{P}-P\subset S$, (hence in W) and $\mathcal{S}$ is contained in some open bounded subset Ω_1 of W and contains only $(0,\lambda_o,t_o)$ as stationary point. Now $\overline{\Omega}_1 \cap S$ is compact and the distance function $d(x,\lambda,t)$ from (x,λ,t) to $\overline{\Omega}_1 \cap S$ is continuous.

Construct an open bounded set Ω such that 1) $\mathcal{S} \subset \Omega \subset \overline{\Omega} \subset \Omega_1$, 2) Ω does not contain stationary points but $(0,\lambda,t)$ for (λ,t) so close to (λ_o,t_o) that $L(\lambda)$ is defined and $(I-\exp tL(\lambda))$ is invertible for $(\lambda,t) \neq (\lambda_o,t_o)$

(for example for (λ,t) in the disc $|\lambda-\lambda_o|^2 + |t-t_o|^2 < \rho_o^2$) and moreover such that 3) $F(x,\lambda,t) \neq 0$ on $\partial\Omega$ except for $(0,\lambda,t)$ with (λ,t) close to (λ_o,t_o). The existence of Ω is an easy consequence of the fact that $\mathcal{S}$ is closed in W and that $S \cap \bar{\Omega}_1$ consists at most of two disjoint compact sets, one being of the form $(0,\lambda,t)$ for (λ,t) close to (λ_o,t_o) since there 0 is an isolated stationary point. (Note that one may have to reduce Ω_1 to a δ-neighborhood of $\mathcal{S}$, with 2δ smaller than the length of the interval Λ_o). Then $\mathcal{S}$ is at a positive distance of the second set in $S \cap \bar{\Omega}_1$. The last property of Ω is obtained as in section two of this chapter.

Consider $H_r(x,\lambda,t) = (F(x,\lambda,t), d(x,\lambda,t)^2 - r^2)$. In a sufficiently small neighborhood of $(0,\lambda_o,t_o)$: $d(x,\lambda,t) = ||x||$ since $(0,\lambda,t)$ is then in S. So for all $r > 0$ $H_r(x,\lambda,t) \neq 0$ on $\partial\Omega$ and for r small enough one can assure that the zeros of $H_r(x,\lambda,t)$ will be in the ball B_o: From the compactness of the solution set in $\bar{\Omega}$ of $F(x,\lambda,t) = 0$ and the construction of Ω, if r is small enough and $H_r(x,\lambda,t) = 0$ then (x,λ,t) is close to (o,λ_o,t_o) and $d(x,\lambda,t) = ||x||$. Choose then ρ such that the disc $\{(o,\lambda,t) \ / \ |\lambda-\lambda_o|^2 + |t-t_o|^2 < \rho^2\}$ is contained in Ω. Then, on the complement of this disc with respect to the intersection of Ω with the plane $x = 0$ near $(0,\lambda_o,t_o)$, one has $||(I - \exp tL(\lambda)^{-1}|| \leq K/2$. So it is enough to choose r so small that $||g(x,\lambda,t)|| \ / \ ||x|| \leq K^{-1}$ for $||x|| \leq r$, to be sure that there are no solutions of H_r outside the ball $B_o = \{(x,\lambda,t) \ / \ ||x||^2 + |\lambda-\lambda_o|^2 + |t-t_o|^2 < \rho^2 + r^2\}$.

Then in order to proceed as in section two, one has to consider an arbitrary space Y of high dimension (may be infinite) and use stable cohomotopy theory by defining $\Sigma H_r(x,\lambda,t) = (F(x,\lambda,t), y, d(x,\lambda,t)^2 + ||y||^2 - r^2)$ on $\Sigma\Omega = \{(x,\lambda,t,y) / (x,\lambda,t) \in \Omega, ||y|| < \tau$ for some $\tau\}$ and then follow the method used above. (If one does not suspend one has trouble defining cohomotopy groups when the dimension n is too small).

Note that here one may use the notion of inessential maps directly: Since on $\partial\Omega$, for r large enough, $H_r(x,\lambda,t)$ will have its component $d(x,\lambda,t)^2 - r^2$ negative ($d(x,\lambda,t)$ is bounded on Ω) then on $M \times \mathbb{R}$ the image of $H_r(x,\lambda,t)$ can be shrunk to a point. This means that $H_r(x,\lambda,t)$ is inessential for all r as noted in Remark 1.3. So choosing a ball B of center $(0,\lambda_o,t_o)$ containing Ω one can extend $H_r(x,\lambda,t)$ to B, for r so small that the only zeros of $H_r(x,\lambda,t)$ are in B_o. Then the extension will be inessential on ∂B while still homotopic to $H_r(x,\lambda,t)|\partial B_o$ via the radial deformation given by $H_r(x,\lambda,t)$ on Ω and its extension to $\overline{B} - \Omega$. ($H_r(x,\lambda,t)$ being not zero in $\overline{B}\text{-}B_o$ this is a valid deformation). This contradicts the fact that on ∂B_o $H_r(x,\lambda,t)$ is essential.

Remark 3.2: Note that the striking fact about Alexander-Yorke's result is that from local properties one is able to infer global consequences. If more is known about the structure of the stationary points then one can get finer results, for example:

Theorem 3.3: Suppose that all bifurcation points in $\overline{P} - P$ are of the form considered in section III of chapter two and have a well defined algebraic multiplicity (in particular they are isolated) then if (x_o,λ_o,t_o) is such a point and $\mathcal{S}$ is the connected component of $\overline{P}$ starting from (x_o,λ_o,t_o), $\mathcal{S}$ is either:

1) <u>Not contained in any compact subset of W</u>

or: 2) $\mathcal{S}$ <u>connects a finite number of stationary points among which there are an even number with odd algebraic multiplicity</u>.

The proof is a straight application of the ideas of section II to $H_r(x,\lambda,t) = (F(x,\lambda,t), d(x,\lambda,t)^2 - r^2)$.

<u>Example 3.4</u>: Suppose that for a discrete set of points D in M and for all λ in Λ, $f(x,\lambda) = 0$ and $f(x,\lambda)$ is otherwise not zero. If $f_x(x,\lambda) = L_x(\lambda)$ is defined for all x in D, continuous in λ and not singular, and if moreover, in case $L_{x_o}(\lambda_o)$ has a pair of pure imaginary eigenvalues, then the corresponding eigenvalues of $L_{x_o}(\lambda)$ for λ near λ_o have non zero real part so that the algebraic multiplicity of this stationary point $(x_o,\lambda_o t_o)$ is well defined. Then since these points are discrete one is in position to apply theorem 3.3.

<u>Remark 3.5</u>: I. Kupka ("closed orbits of one parameter vector fields" preprint) has proved, using transversality, that the situation of theorem 3.1 is generic among C^1 vectors fields on smooth manifolds.

<u>Remark 3.6</u>: It is also possible to give similar results for evolution equations on Banach spaces and in particular for the Navier-Stokes equations. But, since the infinite dimension implies technical difficulties and the use of stronger tools of functional analysis such as semi-groups and Sobolev spaces, this extension is left for a forthcoming article.

CHAPTER FOUR

SPECIAL NONLINEARITIES

After the study of the very small (chapter one) and the very large (chapter three), a middle situation may arise when one assumes some particular features on the nonlinearity $g(x,\lambda)$ in equation

$$(2) \qquad Ax - \lambda x = g(x,\lambda),$$

where A is a Fredholm operator of index zero, having 0 as an isolated eigenvalue (in the sense of section V of chapter one). A is bounded from a Banach space B, continuously imbedded in the Banach space E, into E. Here B and E will be assumed to be real.

Several particular forms of nonlinearities have been studied, using more sophisticated topological tools or algebraic methods. For example assuming that (2), defined on a Hilbert space, is the gradient of some functional f, then the zeros of (2) are the critical points of f and, if f is a Morse function, one can use Morse theory as in Berger [1], or Lusternik-Schnirelman category theory as in Naumann [23], or genus of a set as in Rabinowitz [27], or other variational techniques as in Vainberg [38]. Newton polynomials and algebraic methods have been used by Kirchgässner [21] and Sather [30]. The approach taken in this chapter consists of using simple geometrical facts about a "leading part" which is algebraic and derive from this a topological class which will be invariant under small perturbations. A similar view was taken in Cronin [7], [8], and Dancer [9], [10], where the complex case is treated in great generality.

So, assume that $g(x,\lambda)$ is of the form

$$g(x,\lambda) = N(x,\lambda) + R(x,\lambda)$$

where N and R are continuous mappings from a neighborhood of $(0,0)$ in $B \times \mathbb{R}$ into E. Moreover, retaining the motivation of a Taylor expansion, suppose that there is an integer s, $s > 1$, such that $N(tx, 0) = t^s N(x,0)$ for all positive t and x in B, and $||R(x,\lambda)|| = o(||x||^s)$ for x and λ small. Furthermore assume that there exist r_o, λ_o positive such that for any r, $0 < r < r_o$, one has:

$$(0.1) \qquad ||g(x,\lambda) - g(x',\lambda')||_E \leq Cr^{s-1}(||x-x'||_B + r|\lambda-\lambda'|)$$

$$\text{for } ||x||_B, ||x'||_B \leq r \quad \text{and} \quad |\lambda|, |\lambda'| \leq \lambda_o.$$

This implies that $||g(x,\lambda)||_E \leq C||x||_B^s$ and $g(0,\lambda) = 0$ for all λ, $|\lambda| \leq \lambda_o$, provided $g(0,0) = 0$.

Let Q be the projection on $R(A)$ given by first reducing (2) on $\ker A^\alpha$ as in section III and V of chapter one and using the Jordan form of A on $\ker A^\alpha$, and let K be the pseudoinverse of A. Then for $x = x_1+x_2, x_1$ in $\ker A$, (2) is equivalent to:

$$\begin{cases} (7) \quad x_2 - (I-\lambda KQ)^{-1} \lambda KQx_1 - (I-\lambda KQ)^{-1} KQg(x_1+x_2,\lambda) = 0. \\ (8) \quad \lambda(I-Q)(I-\lambda KQ)^{-1} x_1 + (I-Q)(I-\lambda KQ)^{-1} g(x_1+x_2,\lambda) = 0. \end{cases}$$

Since (0.1) is a stronger condition than (1.6) of chapter one, (7) is uniquely solvable for x_2 as a continuous function of x_1 and λ in a neighborhood of $(0,0)$ in $\ker A \times \mathbb{R}$. Moreover

$$||x_2||_B \leq D_1(|\lambda| \, ||x_1||_B + ||x_1+x_2||_B^s), \quad \text{so}$$

$$||x_2(x_1,\lambda)||_B \leq D(|\lambda|\ ||x_1|| + ||x_1||^s)$$

(and so $||x||_B \leq C||x_1||$ for λ and r_o small enough).

From the choices of Q and P, $\lambda(I-Q)(I-\lambda KQ)^{-1}x_1$ is of the form $\lambda^{k_i} x_1^i$, $i = 1,\ldots,d = \dim \ker A$, and $(I-Q)(I-\lambda KQ)^{-1}g(x_1+x_2,\lambda)$ can be written as $N_o(x_1) + R_o(x_1,\lambda)$ where $N_o(x_1) = (I-Q)N(x_1,0)$ and $R_o(x_1,\lambda) =$

$$(I-Q)R(x_1,0) + (I-Q)(g(x_1+x_2(x_1,\lambda),\lambda) - g(x_1,0))$$

$$+ (I-Q)(I-\lambda KQ)^{-1} \lambda KQg(x_1+x_2(x_1,\lambda),\lambda).$$

So $$||R_o(x_1,\lambda)|| \leq o(||x_1||^s) + C||x_1||^{s-1}(||x_2(x_1,\lambda)||_B + ||x_1||\ |\lambda|) + C|\lambda|\ ||x_1||^s.$$

Hence

$$||R_o(x_1,\lambda)|| \leq K|\lambda|\ ||x_1||^s + o(||x_1||^s) \tag{0.2}$$

for x_1 and λ small enough.

Note that if N is a continuous homogeneous polynomial operator (i.e.: $N(u) = M(u,\ldots,u)$ where M is a s-multilinear form on B), then N satisfies the above conditions. Dancer [9], Lemma 4.

Suppose that the ascent of A, α, is one so that (8) takes the form:

$$\lambda x_1 + N_o(x_1) + R_o(x_1,\lambda) = 0 \tag{0.3}$$

where $R_o(x_1,\lambda)$ satisfies (0.2) and $N_o(tx_1) = t^s N_o(x_1)$. Next, give some inner product, (,), to $\ker A$.

<u>Definition 0.4</u>: u_o, with $||u_o|| = 1$, is called an <u>eigenray</u> of N_o if there exists some real number μ_o such that $N_o(u_o)=\mu_o u_o$.

Then (0.3) can be decomposed in a radial part carried by $x_1 / ||x_1||$ and a tangential part, orthogonal to x_1, defined by the projection P_o from $\mathbb{R}^d$ onto the tangent plane at x_1 to the sphere of radius $r = ||x_1||$, centered at 0. That is:

$$(0.5) \quad \begin{cases} \lambda ||x_1||^2 + (N_o(x_1), x_1) + (R_o(x_1,\lambda), x_1) = 0 \\ P_o N_o(x_1) + P_o R_o(x_1,\lambda) = 0 \end{cases}$$

Suppose that for some u_1, with $||u_1|| = 1$, $P_o N_o(u_1) \neq 0$, then by homogeneity of N_o,

$$||P_o N_o(x_1)|| = ||x_1||^s ||P_o N_o(x_1/||x_1||)|| \geq C ||x_1||^s$$

for any x_1 such that $x_1/||x_1||$ belongs to a small neighborhood of u_1 in S^{d-1}, the unit sphere in ker A, while

$$||P_o R_o(x_1,\lambda)|| \leq K(|\lambda| \, ||x_1||^s + o(||x_1||^s).$$

So for $||x_1||$ and $|\lambda|$ small enough, the tangential part of (0.3) is not zero. Recalling that, if $\mathcal{U}$ is a neighborhood of u_1 in S^{d-1}, one defines a conical neighborhood of u_1 as the set of points tv, for v in $\mathcal{U}$ and $0 < t < 1$, the above argument shows:

Theorem 0.6: <u>Bifurcation for (0.3) cannot occur in conical neighborhoods of points in S^{d-1} which are not eigenrays. And if an eigenray is isolated, bifurcation may occur only in a conical neighborhood of the eigenray.</u>

§ I : STUDY OF EIGENRAYS

Theorem 0.6 suggests looking at continuous maps $N_o(x)$ from $\mathbb{R}^d$ into itself, such that $N_o(tx) = t^s N_o(x)$ for t positive and $s \geq 2$. Consider the restriction of N_o to the unit sphere S^{d-1}. Then, as in (0.5), N_o can be written as the sum of a tangential field $P_oN_o(u)$ and a radial field $\mu(u)\, u$ where $\mu(u) = (N_o(u), u)$.

(15) Assume that for some u_o in S^{d-1}, $P_oN_o(u_o) = 0$ and $\mu(u_o) = \mu_o \neq 0$.

<u>Lemma 1.1</u>: u_o is an isolated zero of P_oN_o in S^{d-1} if and only if u_o is an isolated zero in $\mathbb{R}^d$ of $N_o(x) - \mu_o x$.

<u>Proof</u>: If $P_oN_o(u)$ is not zero in a punctured neighborhood of u_o in S^{d-1} then, by homogeneity of N_o, $N_o(x)$ is not radial for all x, $x \neq 0$, in a cone constructed on the punctured neighborhood. And on the ray u_o, $N_o(x) = ||x||^s\, N_o(u_o) = ||x||^s\, \mu_o u_o = ||x||^{s-1}\, \mu_o x$ is different from $\mu_o x$ for $||x|| \neq 1$, since $s > 1$ and $\mu_o \neq 0$.

The converse is clear since, if in a neighborhood in $\mathbb{R}^d$ of u_o $N_o(x) - \mu_o x$ is zero only at u_o, then $P_oN_o(u) = 0$ for u near u_o (i.e. $N_o(u) = \mu u$) implies that $N_o(x_1) - \mu_o x_1 = \underline{0}$ for $x_1 = u(\mu_o/\mu)^{1/s-1}$ which, by continuity of N_o, is close to u hence to u_o.

So $N_o - \mu_o I$ in $\mathbb{R}^d$ and P_oN_o in S^{d-1} have well defined indices at u_o.

<u>Theorem 1.2</u>: <u>If (15) holds, then</u>:

Index $(N_o - \mu_o I, u_o) = \text{Sign}\, \mu_o \cdot \text{Index}\, (P_oN_o, u_o)$.

<u>Proof</u>: Let B_o be a small ball imbedded in S^{d-1} around u_o and let B be the truncated conical neighborhood of u_o in $\mathbb{R}^d$ defined as:

$$B = \{x \in \mathbb{R}^d \ / \ \tfrac{1}{2} < ||x|| < 3/2, \ x/||x|| \in B_o\}.$$

Then

$N_o(x) - \mu_o x = P_o N_o(x/||x||)||x||^s \oplus \mu(x/||x||)||x||^s - \mu_o||x||$

by taking at each point the tangential and normal part of

$N_o(x) - \mu_o x$.

On the boundary ∂B of B, this field can be deformed to

$P_o N_o(x/||x||) \oplus \mu_o(||x||^s - ||x||)$ via

$P_o N_o(x/||x||)||x||^{s(1-t)} \oplus ((1-t)\mu(x/||x||)+t\mu_o)||x||^s - \mu_o||x||$,

since, if for some x on ∂B this deformation is zero, then $N_o(x/||x||)$ is normal and by choice of B_o: $x/||x|| = u_o$, so that $\mu(x/||x||) = \mu_o$ and the second part cannot be zero for $||x|| = 1/2$ or $3/2$. So, by the product theorem:

degree $(N_o(x)-\mu_o x, B, 0)$ = degree $(P_o N_o(x/||x||), B_o, 0) \times$ degree $(\mu_o(r^s-r),]1/2, 3/2[\ , 0)$. The second degree is easily seen to be sign $((s-1)\mu_o)$ = sign μ_o (since $s > 1$) by taking its derivative at $r = 1$, hence the proof of the theorem.

<u>Theorem 1.3</u>: <u>Suppose that</u> $N_o(x)$ <u>is the gradient of</u> $\ell(x) = (N_o(x), x)/s+1$. <u>If</u> u_o <u>in</u> S^{d-1} <u>is an isolated, relative extremun of</u> $\ell(u)$ <u>for</u> u <u>in</u> S^{d-1}, <u>with</u> $\ell(u_o) \neq 0$, <u>Then</u>: Index $(P_o N_o, u_o) = (-1)^{d-1}$ <u>for a maximum</u>, 1 <u>for a minimum</u>.

<u>Remark 1.4</u>: If $N_o(x)$ is a polynomial map, the fact that $\ell(x)$ has this form is a consequence of the Euler Identity. Sather [30], p. 237.

<u>Proof of the theorem</u>: Let u_o in S^{d-1} be a critical point of ℓ restricted to S^{d-1}, i.e. $N_o(u_o) = \mu_o u_o$ where $\mu_o = (N_o(u_o), u_o)$ is a Lagrange multiplier. Then u_o is an isolated critical point of ℓ if and only if u_o is an isolated

zero of the tangential field P_oN_o. So, under the hypothesis of the theorem, index (P_oN_o, u_o) is well defined. Choose B_o, a ball in S^{d-1} around u_o, so that it contains no other critical point of ℓ. For a maximum (resp. minimum) let $\varepsilon = \ell(u_o) - \sup_{u \in \partial B_o} \ell(u)$, (resp. $\varepsilon = \inf_{u \in \partial B_o} \ell(u) - \ell(u_o)$),

and let M be $\ell^{-1}(]\ell(u_o) - \varepsilon, \ell(u_o)[) \cap B_o$

(resp. $M = \ell^{-1}(]\ell(u_o), \ell(u_o) + \varepsilon[) \cap B_o$). Then M is a smooth open submanifold of B_o with a smooth boundary $\ell^{-1}(\ell(u_o) - \varepsilon))$ (resp. $\ell^{-1}(\ell(u_o) + \varepsilon)$ and an isolated boundary point u_o, since any value between $\ell(u_o)$ and $\ell(u_o) \pm \varepsilon$ is regular for ℓ restricted to B_o. Moreover P_oN_o is normal to the level curve $\ell(u_o) \pm \varepsilon$, inward for a maximum, outward for a minimum. Finally, the Poincaré-Hopf theorem implies: Index (P_oN_o, u_o) = Lefschetz number of $P_oN_o = (-1)^{d-1}\chi(M)$ for an inward field and $\chi(M)$ for an outward field, where $\chi(M)$, the Euler characteristic of M, is $\sum_0^{d-1} (-1)^i \dim H_i(M;Q)$. Then a contraction of M to u_o, along the field P_oN_o, permits us to conclude that all homology groups are zero except for i equals zero, thus giving the index of P_oN_o.

<u>Remark 1.5</u>: If the level curves $\ell^{-1}(\ell(u_o) \pm \varepsilon)$ are $(n-2)$ spheres, then an inward field is homotopic to $-\mathrm{Id}$ while an outward field is homotopicto Id, and it is well known that the index of $-\mathrm{Id}$ on S^{d-2} is $(-1)^{d-1}$.

<u>Remark 1.6</u>: Rothe in [28], [29] has shown that, under certain conditions on P_oN_o (satisfied for extrema), the index of P_oN_o at a critical point is $\sum_0^{d-1} (-1)^r M_r$, where M_r is the r-th Morse type number of the critical point, so that theorem 1.3 can be extended to this case.

Remark 1.7: The above analysis can be used to simplify and extend the following result of A. Grundmann (D.D.N. Thesis Stuttgart 1974).

Let F and A be two homogeneous polynomial maps, from $\mathbb{R}^d$ to $\mathbb{R}^d$, of respective degree β and α, which are the gradients of $\phi(x) = \frac{1}{1+\beta} \langle F(x),x\rangle$ and $\psi(x) = \frac{1}{1+\alpha} \langle A(x),x\rangle$. Assume there exists $x_o \neq o$ such that: $A(x_o) = F(x_o)$, $\psi(x_o) > o$ and x_o is an isolated local extremum of $\phi(x)$ for x on the surface $N = \{x \in \mathbb{R}^d \ / \ \psi(x) = \psi(x_o)\}$. One has then the following result:

If $\beta > \alpha$, the index of $F-A$ in $\mathbb{R}^d$ is well defined and equal to $(-1)^{d-1}$ for a maximum of ϕ and 1 for a minimum.

In fact, the condition $\psi(x_o) > o$ insures that N is a smooth hypersurface in $\mathbb{R}^d$ with normal $A(x_o)$ at x_o. So, from a tangential and normal decomposition of F (replacing S^{d-1} by N), it is easy to see that, $(\beta \neq \alpha)$, as in Lemma 1.1 the tangential part of F, PF, has an isolated zero i.e. a critical point of ϕ on N, if and only if $F-A$ has an isolated zero in $\mathbb{R}^d$ and, as in theorem 1.2, the index of $F-A$ in $\mathbb{R}^d$ is Index $(N,PF) \times \text{sign}\ (\beta-\alpha)$. The computation of this index then follows as in theorem 1.2.

§ II : APPLICATION: BIFURCATION NEAR AN ISOLATED EIGENRAY

Theorems 1.2 and 1.3 will be applied to equation

$$\lambda x_1 + N_o(x_1) + R_o(x_1,\lambda) = 0 \tag{0.3}$$

where $||R_o(x_1,\lambda)|| \leq K|\lambda|\ ||x_1||^s + o(||x_1||^s)$.

Definition 2.1: Let u_o in S^{d-1} be an eigenray with $N_o(u_o) = \mu_o u_o$, $\mu_o \neq 0$. Then (0.3) has a bifurcating branch in the direction u_o if there exists λ_o such that, for $0 < |\lambda| < \lambda_o$, (0.3) has a solution $x_1(\lambda) \neq 0$, with $||x_1(\lambda)|| \to 0$, as $|\lambda| \to 0$ and $x_1(\lambda)\ /\ ||x_1(\lambda)|| \to u_o$, as $\lambda \to 0$. If the above is true only for positive λ, then one has bifurcation above 0 and, if only for negative λ, bifurcation below 0.

Theorem 2.2: Suppose that N_o is the gradient of $\ell(u) = (N_o(u),u)/s+1$. Let u_o be an isolated eigenray of N_o, corresponding to a relative extremum $\mu_o/s+1$, of $\ell(u)$ restricted to S^{d-1}. Then:

(a): (0.3) bifurcates below 0 in the direction u_o if $\mu_o > 0$,

(b): (0.3) bifurcates above 0 in the direction u_o if $\mu_o < 0$.

Remark 2.3: If N_o is even on S^{d-1} then $N_o(u_o) = \mu_o u_o$ implies $N_o(-u_o) = -\mu_o(-u_o)$ and, if μ_o is a local maximum (minimum) of ℓ at u_o, $-\mu_o$ will be a local minimum (maximum) of ℓ at $-u_o$, so that in this case bifurcation occurs above and below 0, one branch corresponding to the direction u_o, the other to the direction $-u_o$.

If N_o is odd, then $N_o(\pm u_o) = \mu_o(\pm u_o)$, so that one gets bifurcation above or below 0, but in the two directions $\pm u_o$.

Proof of the theorem 2.2: (a). Suppose that μ_o is positive and $N_o(u_o) - \mu_o u_o = 0$. For λ negative, let $x_1 = |\lambda/\mu_o|^{1/s-1}\ x$. Then using homogeneity of N_o, (0.3) becomes:

$$- |\lambda|\ |\lambda/\mu_o|^{1/s-1}\ x + |\lambda/\mu_o|^{s/s-1}\ N_o(x) + R_o(|\lambda/\mu_o|^{1/s-1}\ x,\lambda)$$

and multiplying by $|\mu_o/\lambda|^{s/s-1}$:

$$(2.4) \qquad -\mu_o x + N_o(x) + R_1(x,\lambda) = 0$$

where $R_1(x,\lambda) = |\lambda/\mu_o|^{-s/s-1} R_o(|\lambda/\mu_o|^{1/s-1} x,\lambda)$ satisfies, (from (0.2))

$$||R_1(x,\lambda)|| \leq K|\lambda| \; ||x||^s + |\lambda/\mu_o|^{-s/s-1} \; o(|\lambda/\mu_o|^{s/s-1}||x||^s),$$

so that for $||x||$ bounded, $||R_1(x,\lambda)||$ goes uniformly to 0 when λ tends to 0.

From theorems 1.3 and 1.2, index $(-\mu_o I + N_o, u_o) = \pm 1$. So choosing a ball B in $\mathbb{R}^d$, around u_o, degree $(-\mu_o x + N_o(x) + R_1(x,\lambda), B, 0)$ is defined for $|\lambda|$ sufficiently small, $0 < |\lambda| < \lambda_o$, giving a solution $x(\lambda)$ to (2.4) for each λ, with $x(\lambda)$ tending to u_o when λ goes to 0. So $x_1(\lambda) = |\lambda/\mu_o|^{1/s-1} x(\lambda)$ tends in norm to 0 and in direction to u_o, proving part (a).

(b) If μ_o is negative and $N_o(u_o) + |\mu_o|u_o = 0$, then, for positive λ, let $x_1 = |\lambda/\mu_o|^{1/s-1} x$, so that (0.3) reduces to $|\mu_o|x + N_o(x) + R_1(x,\lambda)$ for which the analysis is similar.

<u>Remark 2.5</u>: Here the fact that N_o was the gradient of $\ell(u)$ was used only to compute the degree of (0.3) near u_o. If N_o has an isolated eigenray u_o such that Index $(P_o N_o, u_o)$ is not zero, the application of theorem 1.2 and the arguments of theorem 2.2 will give the same type of bifurcation.

As a consequence of theorem 2.2, one obtains theorems 4.6 and 4.7 of Sather [30]. Namely:

<u>Corollary 2.6</u>: <u>If moreover</u>, $d = 2$ <u>and</u> N_o <u>is a polynomial map, then any eigenray is isolated provided that, for</u> s <u>even</u>

N_o <u>is not identically zero, for s odd</u> $(N_o(u),u)$ <u>is not constant on S^1</u>.

<u>Proof</u>: Let u_o be in S^1 such that $P_oN(u_o) = 0$, i.e. if u_1 is a unit vector orthogonal to u_o, then $(N_o(u_o),u_1) = 0$. If u_o is not an isolated zero of P_oN_o, there exists a sequence α_n,α_n tending to 1 as n tends to ∞, and β_n such that: $\alpha_n^2 + \beta_n^2 = 1$ and $(N_o(\alpha_n u_o+\beta_n u_1),\beta_n u_o-\alpha_n u_1) = P_oN_o(\alpha_n u_o+\beta_n u_1) = 0$.

Let t_n be β_n/α_n, then t_n tends to zero and, by homogeneity of N_o : $(N_o(u_o+t_n u_1),t_n u_o-u_1) = 0$. Since N_o is a polynomial map, this implies that $(N_o(u_o+tu_1), tu_o-u_1) \equiv 0$ for all t in $\mathbb{R}$. So $N_o(u) = \mu(u)u$ for u in S^1, using the fact that $N_o(u) = \pm N_o(-u)$. Hence $N_o(x) = \mu(x/||x||)||x||^{s-1}x$.

Let h be such that $u_o + h = u_o \cos\theta + u_1 \sin\theta$, so $||u_o+h|| = 1$ and $\mu(u_o+h)$ is a smooth function $\mu(\theta)$ of θ. Then:

$$(2.7)\quad (N_o(u_o+h),u_o+h)-(N_o(u_o),u_o) = (s+1)(N_o(u_o),h) + r(u_o,h)$$

with $r(u_o,h) = o(||h||)$, by definition of a gradient. Now since $h = u_o(1-\cos\theta) + u_1\sin\theta$, $||h|| = (2(1-\cos\theta))^{\frac{1}{2}}$ and $||h||/|\theta|$ tends to 1 as θ tends to 0, so $r(u_o,h) = o(|\theta|)$.

Equation (2.7) becomes $\mu(\theta)-\mu(0) = (s+1)\mu(0)(1-\cos\theta) + o(|\theta|)$ which implies that: $\left.\frac{d\mu(\theta)}{d\theta}\right|_{\theta=0} = 0.$

Since this can be done at any u_o in S^1, then $\mu(\frac{x}{||x||})$ is constant and $N_o(x) = \mu_o||x||^{s-1}x$ which is not a polynomial if s is even and μ_o is not zero, while if s is odd

$(N_o(u),u) = \mu_o$ contradicts the hypothesis of the corollary.

Remark 2.8: In [21], Kirchgässner, using implicit function theorem and the Brouwer fixed point theorem, has established a result which seems stronger than theorem 2.2 and corollary 2.6, in the sense that the critical point u_o needs not be isolated. However, part of his proof relies on Remark 2 in [21] which asserts that if s is even, then a critical point corresponding to a minimum of $\ell(u)$ restricted to S^{d-1} is isolated. But consider in $\mathbb{R}^3$:

$$\ell(x,y,z) = \sqrt{2}\ /5\ z\,[(x^2+y^2-z^2)^2 + 5/4(x^2+y^2+z^2)^2 - z^4],$$

then N_o has for components N_x, N_y, N_z with:

$$N_x = 4/5\ \sqrt{2}\ xz\ [(x^2+y^2-z^2) + 5/4(x^2+y^2+z^2)] \quad ,$$

$$N_y = 4/5\ \sqrt{2}\ yz\ [(x^2+y^2-z^2) + 5/4(x^2+y^2+z^2)] \quad ,$$

$$N_z = 4/5\ \sqrt{2}\ z^2\ [(z^2-x^2-y^2) + 5/4(x^2+y^2+z^2)-z^2] +$$

$$\sqrt{2}\ /5[(x^2+y^2-z^2)^2 + 5/4(x^2+y^2+z^2)^2-z^4]$$

Here $s = 4$. Now, on the set A of points (x,y,z) such that $x^2 + y^2 + z^2 = 1$, $x^2 + y^2 = z^2$, $z = 1/\sqrt{2}$, then $N_x(x,y,z) = x$, $N_y(x,y,z) = y$, $N_z(x,y,z) = z = 1/\sqrt{2}$; so any point in A is a fixed point of N_o, hence a critical point of ℓ. Moreover $\ell(u) = 1/5$ for any u in A and ℓ restricted to S^2 is minimum on A: In fact if $x^2 + y^2 = \frac{1+\varepsilon}{2}$, $z^2 = \frac{1-\varepsilon}{2}$ then

$$\ell(u) = 1/5\ \sqrt{1-\varepsilon}\ (1 + \varepsilon/2 + 3\varepsilon^2/4) =$$

$$1/5(1 - \varepsilon/2 - \varepsilon^2/8 + \ldots)(1 + \varepsilon/2 + 3\varepsilon^2/4) = 1/5(1 + 3\varepsilon^2/8 + \ldots)$$

In addition, when $d = 2$, the corollary to theorem 2, in [21], says that for $s \geq 3$, u_o corresponding to an extremum of ℓ on S^1 and $R_o(x,\lambda) = \lambda S(x,\lambda)$ such that $(S(u_o,0),u_o) \neq 0$, then:

$$(2.9) \qquad x - N_o(x) + \lambda S(x,\lambda) = 0$$

has a solution in a neighborhood of u_o, for any small non-zero λ. But suppose that: $N_x(x,y) = x(x^2+y^2)^k$, $N_y = y(x^2+y^2)^k$ so $\ell(x,y) = (x^2+y^2)^{k+1}/2(k+1)$ and $s = 2k+1$. Assume also that: $S(x,\lambda) = (x(x^2+y^2)^{2k},0)$. So if one excludes neighborhoods of the four points $(\pm 1,0)$, $(0,\pm 1)$, (2.9) has no solution, for λ not zero, near any other point u_o of S^1, although $(S(u_o,0),u_o) = x^2$ is not zero for these points. (One needs $y = 0$ or $x^2 + y^2 = 1$ in the second equation and, $y = 0$ being excluded, this implies $x = 0$ or $\lambda = 0$ which is not possible).

A last case will be studied: Suppose that N_o is not a gradient but is a homogeneous polynomial map of degree s. Moreover assume that d is 2. Then for any couple u_1, u_2 of orthogonal unit vectors in $\mathbb{R}^2$, define

$$Q_{u_1}(t) = (N_o(u_1+tu_2), tu_1-u_2) = (1+t^2)^{\frac{s+1}{2}} P_o N_o((u_1+tu_2)/||u_1+tu_2||)$$

so that a zero of $Q_{u_1}(t)$ corresponds to an eigenray u_o of N_o, if $N_o(u_o)$ is not zero.

Theorem 2.10: (Sather [30], Theorem 4.5 and corollary.) *If for some u_1 in S^1, $Q_{u_1}(t)$ has a root at $t = 0$ of odd multiplicity, then (0.3) has a branching in the direction u_1 above 0 $(\lambda > 0)$ if $(N_o(u_1),u_1) < 0$, below 0 $(\lambda < 0)$ if $(N_o(u_1),u_1) > 0$.*

Remark 2.11: Note that if $Q_{u_1}(t)$ is not identically zero, then the eigenrays are isolated (as in the first part of the proof of corollary 2.6). Only in this case the multiplicity of a zero of $Q_{u_1}(t)$ is defined.

Proof of theorem 2.10: For $\lambda > 0$ and $(N_o(u_1),u_1) = -|\mu_1|$, as in theorem 2.2 reduce (0.3) to $|\mu_1|x + N_o(x) + R_1(x,\lambda) = 0$ Since at $t = 0$, $Q_{u_1}(t)$ changes sign, so does $P_oN_o(u_1+tu_2)/||u_1+tu_2||)$ i.e. P_oN_o has index ± 1 near u_1. Then theorem 1.2 and the estimation of theorem 2.2 gives the result.

Corollary 2.12: *If moreover s is even and $||N_o|| \neq 0$ on S^1, then there exists u_1 (and $-u_1$) giving branchings above and below* 0.

Proof: It is enough to show that $Q_{u_1}(t)$ is a polynomial of odd degree $s + 1$ (then necessarily $Q_{u_1}(t)$ will have a real root of odd multiplicity) and use Remark 2.3.

Now the leading coefficient of $Q_{u_1}(t)$ is $(N_o(u_2),u_1)$. If it is zero for all choices of u_1, then P_oN_o is identically zero and $N_o(u) = \mu(u)u$ where $\mu(u)$ is an odd mapping from S^1 to $\mathbb{R}$ (s is even). By the Borsuk Antipodensatz this mapping must have a zero on S^1, which contradicts the nondegeneracy of N_o (Granas [16], Appendix).

Remark 2.13: In the case $d = 2$, one can analyse in more detail the remainder $R_o(x,\lambda)$. See, for example, Dancer [9], Theorem 8.

Remark 2.14: If the ascent of A is not one, then one can still give interesting bifurcation results, but not in the direction of an eigenray. In this case (8) reduces to:

$$(2.15)\qquad \lambda^{k_i} x_1^{(i)} + N_o^{(i)}(x_1) = R_o^{(i)}(x_1,\lambda) = 0 \qquad i = 1,\dots,d.$$

Suppose that N_o is a homogeneous polynomial of degree s, vanishing only at 0. Then for λ small enough, but fixed, the dominant term in (8) will be, for $||x_1|| = \rho$ very small, $\lambda^{k_i} x_1^{(i)}$ $i = 1,\dots,d$ and for $||x_1|| = R$ larger, N_o. So that if degree $((2.15), ||x_1|| < R, 0)$ is different from the index of $\lambda^{k_i} x_1^{(i)}$ at 0, (λ fixed), then one has a solution x, with $\rho < ||x_1|| < R$. The first degree is the index of N_o and is even if and only if s is even (Cronin [6], p. 49), while the second is ± 1: Dancer [9], theorem 9. In the complex case, the index of N_o is sd, so there is always bifurcation: Sather [30], theorem 3.4. See also Dancer [10].

Remark 2.16: In the above situation, suppose that $k_1 \leq k_2 \leq \dots \leq k_d$ and that $k_1 = k_2 = \dots = k_r < k_{r+1}$ for some r. For λ positive divide (2.5) by $\lambda^{k_1 s/s-1}$ and let $x = x_1 \lambda^{-k_1/s-1}$. So, one gets:

$$(2.17)\qquad \lambda^{k_i - k_1} x^{(i)} + N_o^{(i)}(x) + R_1(x,\lambda)\ ,$$

where $R_1(x,\lambda) = \lambda^{-k_1 s/s-1} R_o(x\lambda^{k_1/s-1},\lambda)$ satisfies

$$||R_1(x,\lambda)|| \leq K|\lambda|\ ||x||^s + |\lambda|^{-k_1 s/s-1} o(||x||^s|\lambda|^{k_1 s/s-1})$$

If s is odd, one can use also this reduction for negative λ, so (2.17) and $\sum_1^d |x^{(i)}|^2 - \rho^2$ will have a degree -2, if $\sum_{i=1}^{d} (k_i - k_1)$ is odd, as was seen in theorem 3.2 and corollary

3.8 of chapter one, that is if $m - dk_1$ is odd, where m is the multiplicity of 0 and d the dimension of $\ker A$.

If moreover:

(2.18) $\sum_{r+1}^{d} |N_o^{(i)}(x)|^2$ is zero only for $x = 0$

Then (2.17) cannot have its solution for $||x|| = \rho$ and $\lambda = 0$, since then $R_1(x,0) = 0$ and $N_o^{(i)}(x) = 0$ for $i = r+1,\ldots,d$ which implies $x = 0$. So under the above condition (2.18) and s odd, the original equation (2.15) will have a bifurcating solution.

Also if s is even, λ fixed small and $\sum_{r+1}^{d} |N_o^{(i)}(x)|^2$ nondegenerate, then the product theorem, for $\sum_{1}^{r} |x^{(i)}|^2$ small and $\sum_{r+1}^{d} |x^{(i)}|^2 = \rho$ or R, will give, in one case, an index 1 (for ρ so small that $\lambda^{k_i - k_1} x^{(i)}$ dominate (2.17)) and, in the other case, a degree congruent (modulo two) to $s(d-r)$, hence even, for R so large that $x^{(1)},\ldots,x^{(r)}$ and $N_o^{(r+1)}(0,\ldots,0,x^{(r+1)},\ldots,x^{(d)}),\ldots,N_o^d(0,\ldots,0,x^{(r+1)},\ldots,x^{(d)})$ dominate (2.17). $((N_o(x) - N_o(0,\ldots,0,x^{(r+1)},\ldots,x^{(d)})$ is small compared to $N_o(0,\ldots,0,x^{(r+1)},\ldots,x^{(d)})$ for a judicious choice of R). Since this solution of (2.17) is for λ not zero (a similar argument works for negative λ), it gives rise to a solution of (2.15): Dancer [9], Theorem 10.

Remark 2.19: Global results can be derived as well for the type of nonlinearities studied in this chapter: In fact, consider the equation:

$$(2.20) \qquad (I-\lambda T)x - N(x) - R(x,\lambda) = 0$$

where T is a compact linear operator on a real Banach space E, N and R are compact mappings satisfying condition (0.1) and N is homogeneous of degree s.

Assume the following:

1) λ_o is a characteristic value of ascent one. (So locally (2.20) has the form (0.3).).

2) u_o is an isolated eigenray in $\ker(I-\lambda_o T)$ along which (2.20) bifurcates.

3) Let $\mathcal{E}$ be the connected component of the set P of non-trivial solutions to (2.20) which contains the branch bifurcating in the direction u_o. Then $\mathcal{E}$ is bounded and whenever $\overline{\mathcal{E}}$ meets $(0,\lambda_i)$ (λ_i characteristic value) it does so along an isolated eigenray u_i^j in $\ker(I-\lambda_i T)$ and λ_i has ascent one.

Given the decomposition of E as the direct sum of the kernel and range of $I-\lambda_i T$, let N_i denote the projection on N restricted to $\ker(I-\lambda_i T)$ onto that kernel. Moreover, if u is an element of the unit sphere S^{d_i-1} in the kernel, $P_o N_i(u)$ will stand for the tangential part of $N_i(u)$. One has then the following result:

Theorem 2.21: Let β_i be the sum of the multiplicities of $I-\mu\lambda_i T$ for $0 < \mu \leq 1$. Then:

$$\sum_{i,j} (-1)^{\beta_i} \operatorname{sign} \lambda_i \operatorname{Index} (P_o N_i ; u_i^j) = 0$$

Note that $(0,\lambda_i)$ is directly connected to $(0,\lambda_o)$ by $\overline{\mathcal{E}}$ and does not belong to $\mathcal{E}$; this is a sharper result than the one given in chapter three where this connection could be done via another characteristic value in $\overline{\mathcal{E}}$.

Note also that the same reasoning could be extended to the case of a complex parameter as well as for several parameters with the appropriate hypothesis. (In particular one has to consider isolated components of eigenrays instead of points and translate the sum into stable cohomotopy classes. This will give a correct proof to the paper of R.J. Magnus, "A homotopy approach to global branching problems" Battelle-Geneva math. report,(1975)).

Proof of the theorem: The result will follow from the computation of a local index: Let P be the projection on $\ker(I-\lambda_o T)$ giving the classical decomposition of E, then any x in E is equal to $x_1 + x_2$ with x_1 in the kernel and x_2 in the range of $I-\lambda_o T$. (2.20) can be written as:

$$\begin{cases} (I-\lambda T)x_2 - (I-P)\{N(x) + R(x,\lambda)\} = 0 \\ (1-\lambda/\lambda_o)x_1 - P\{N(x) + R(x,\lambda)\} = 0 \end{cases}$$

Then as usual the first equation is uniquely solvable for x_2 in terms of x_1 and λ with the estimate:

$$||x_2|| \le D(|\lambda-\lambda_o|\ ||x_1|| + ||x_1||^s)$$

So, on the sphere $||x||^2 + |\lambda-\lambda_o|^2 = r^2$, one has $||x_2|| \le C\, r^2$.

As for (0.3), the second equation is transformed into:

$$(1-\lambda/\lambda_o)x_1 - N_o(x_1) - R_o(x_1,\lambda) = 0,$$

and the known part $(R_o = 0)$ has then the tangential-radial decomposition:

$$\begin{cases} -P_oN_o(x_1) = 0 \\ (1-\lambda/\lambda_o)||x_1||^2 - (N_o(x_1),x_1) = 0 \end{cases}$$

So, if $N_o(u_o) = \mu_o u_o$ and since u_o is an isolated eigenray, one can find a conical neighborhood of u_o so that the tangential part will have solutions of the form ρu_o and the radial part of the form $\lambda = \lambda_o - \lambda_o \mu_o \rho^{s-1}$. This means that the eigenray ρu_o in $\ker(I-\lambda_o T)$ corresponds to a curved line:

$$(x_2(\rho u_o, \lambda_o - \lambda_o \mu_o \rho^{s-1}), \rho u_o, \lambda_o - \lambda_o \mu_o \rho^{s-1}) \quad \rho > 0,$$

which will cut the sphere $||x||^2 + |\lambda - \lambda_o|^2 = r^2$ in just one point, say for ρ_o. Let B_o be a small ball on this sphere and around that intersection point A, construct then a truncated cone on B_o and along the above curved line. On the boundary of this cone, the pair (2.20) together with the side condition $||x||^2 + |\lambda - \lambda_o|^2 - r^2 = 0$ has, for r small enough, no solution and can be deformed to:

$$((I-\lambda T)x_2 - (1-\lambda/\lambda_o)x_1 - N_o(x_1), ||x_1||^2 + |\lambda-\lambda_o|^2 - r^2)$$

which has an isolated zero at A and so a well defined index. This index is the product of $(-1)^{\beta_o - d}$ by the index of the two last terms at the point $(\rho_o u_o, \lambda_o - \lambda_o \mu_o \rho_o^{s-1})$. (Here d denotes the dimension of $\ker(I-\lambda_o T)$ and $(-1)^{\beta_o - d}$ represents the index of $(I-\lambda T)x_2$ at 0; Recall that λ_o has ascent one).

As in the proof of theorem 1.2, this last index is the product of $\mathrm{Index}(-P_o N_o(u)\big|_{S^{d-1}}; u_o)$ by the index of

$$\{(1-\lambda/\lambda_o)\rho^2 - \mu_o \rho^{s+1}, \rho^2 + |\lambda - \lambda_o|^2 - r^2\} \text{ at } (\rho_o, \lambda_o - \mu_o \lambda_o \rho_o^{s-1}),$$

where ρ_o is the unique positive solution of the expression $\rho^2 + \lambda_o^2 \mu_o^2 \rho^{2(s-1)} - r^2 = 0$ an increasing function of ρ.

The first index is: $(-1)^d \mathrm{Index}(P_o N_o; u_o)$ and the second is the sign of the Jacobian with respect to ρ and λ, that is: $\mathrm{sign}\,(2\mu_o^2(s-1)\lambda_o \rho_o^{2s-1} + 2\rho_o^3/\lambda_o) = \mathrm{sign}\,\lambda_o$.
So the degree of (2.20) and its side condition on the truncated

curved cone is: $(-1)^{\beta_o}$ sign λ_o Index $(P_oN_o;u_o)$.

Construct then an open bounded set Ω in $E \times \mathbb{R}$ such that:

1) $\mathcal{E}$ is contained in Ω.

2) If (x,λ) in $\mathcal{E}$ is close to $(0,\lambda_i)$ and in the direction u_i^j then Ω is locally a curved conical neighborhood constructed around u_i^j as above, (this is possible according to theorem 0.6). Note that $(0,\lambda_i)$ belongs to $\partial\Omega$.

3) (2.20) is not zero on $\partial\Omega$ except at $(0,\lambda_i)$.

Then $H_r(x,\lambda) = ((2.20),\ d(x,\lambda;S)^2 - r^2)$ is not zero on $\partial\Omega$, for all positive r. Here S denotes the set of points $(0,\lambda_i)$ and $d(\cdot,\cdot;\cdot)$ is the distance function. Since Ω is bounded, for r very large H_r will be not zero on Ω and so: degree $(H_r,\Omega,0) = 0$. Moreover, if (x,λ) belongs to the compact set $\overline{P} \cap (\overline{\Omega} - \cup B_i)$ where B_i are small balls around $(0,\lambda_i)$ so that $\Omega \cap B_i$ is the union of conical neighborhoods of u_i^j, then one has a positive lower bound on $d(x,\lambda;S)$. So, for r small enough, a zero of H_r must be in one of these conical neighborhoods where $d(x,\lambda;S) = ||x||^2 + |\lambda-\lambda_i|^2$. This implies that: degree $(H_r,\Omega,0)$ is the sum of the local indices computed in the first part of the proof.

APPENDIX

In order to make this paper as self contained as possible, a complete proof of lemma 3.5 of chapter two is included here.

Recall that one wants to compute the homotopy class of the map $I\text{-}\exp tL(\lambda)$ from $B^2(\rho,r) = \{(\lambda,t) \ / \ |\lambda-\lambda_o| \leq \rho, |t-t_o| \leq r\}$ to the set $M(n)$ of real $n \times n$ matrices. $L(\lambda)$ has the following properties: It is an invertible matrix with distinguished eigenvalues $\mu_j(\lambda) = \alpha_j(\lambda) \pm i\,\beta_j(\lambda)$, $j = 1,\ldots,k$ such that: $\alpha_j(\lambda_o) = 0, \beta_j(\lambda_o) = k_j\beta$ with k_j positive integers $(k_1 = 1)$ and β a real number defining $t_o = 2\pi/\beta$. If $\alpha_j(\lambda) \neq 0$ for $\lambda \neq \lambda_o$ then $I - \exp tL(\lambda)$ is invertible for all (λ,t) in $B^2(\rho,r)$ except for (λ_o,t_o). So, assuming that the determinant of the above matrix is positive on the boundary $S^1(\rho,r)$ of $B^2(\rho,r)$, this matrix represents an element of $\pi_1(GL(n)^+)$.

Lemma 3.5 states that this element is non-trivial if and only if an odd number of the $\alpha_j(\lambda)$ change sign as λ passes through λ_o.

The proof given here is not very different from the original proof of Alexander and Yorke ([0] Lemma 8.1). In particular the author has been unable to avoid the use of the M-Structure Lemma.

From Kato ([20] Theorems II-5-1 and II-5-2) one knows that:

a) $\mu_j(\lambda)$ can be numbered in such a way that they are continuous in λ.

b) the projection $P(\lambda)$ corresponding to the totality of the $\mu_j(\lambda)$ and $\overline{\mu}_j(\lambda)$ is continuous in λ. So for each λ, one has: $\mathbb{R}^n = X_o(\lambda) \oplus X_1(\lambda)$, $P(\lambda)$ being the projection on $X_o(\lambda)$, furthermore both subspaces are invariant under $L(\lambda)$ and the dimensions of $X_i(\lambda)(i=0,1)$ are constant for all λ near λ_o.

c) Choosing a basis such that $L(\lambda_o)$ is in the block form given below, then, for λ close to λ_o, there exists an isomorphism $F(\lambda)$ continuous in λ and close to the identity such that $F(\lambda)^{-1}L(\lambda)F(\lambda)$ is represented by:

$$\begin{bmatrix} B(\lambda) & 0 \\ 0 & \tilde{B}(\lambda) \end{bmatrix}$$

$B(\lambda)$ corresponding to $P(\lambda)$ and $\tilde{B}(\lambda)$ to the other eigenvalues.

Consider the following matrix which is similar to $I\text{-}\exp tL(\lambda)$ and as such invertible on $S^1(\rho,r)$:

$$I\text{-}\exp t\{(1-\tau)I + \tau F(\lambda)\}^{-1}L(\lambda)\{(1-\tau)I + \tau F(\lambda)\}$$

(One may have to reduce ρ so that $F(\lambda)$ is so close to I that $(1-\tau)I + \tau F(\lambda)$ is invertible. Note that the choice of basis for $L(\lambda_o)$ will at most change the sign of the class).

So $I\text{-}\exp tL(\lambda)$ is homotopic to:

$$\begin{bmatrix} I\text{-}\exp tB(\lambda) & 0 \\ 0 & I\text{-}\exp t\tilde{B}(\lambda) \end{bmatrix}$$

Now, since $I\text{-}\exp t\tilde{B}(\lambda)$ is invertible in $B^2(\rho,r)$, one may deform $\tilde{B}(\lambda)$ to $\tilde{B}(\lambda_o)$, via $B((1-\tau)\lambda + \tau\lambda_o)$ and $I\text{-}\exp t\tilde{B}(\lambda_o)$ to I_{n-2k} or to $I_{n-2k-1} \oplus -I_1$ according to the component of $GL(n-2k)$ to which it belongs. Assuming that $I\text{-}\exp t\tilde{B}(\lambda_o)$ has positive determinant, then the class of

$I-\exp tL(\lambda)$ in $\pi_1(GL(n)^+)$ is the "suspension" of the class of $I-\exp tB(\lambda)$ in $\pi_1(GL(2k)^+)$. (In the other case one would have to multiply by a map of degree -1).

One may further perform this diagonalization process for each group of μ_j corresponding to different k_j, but in general one cannot separate the eigenvalues splitting from $ik_j\beta$ so one has to use the following detour:

Let C_1 be the subset of $M(2k)$ of matrices with at least one pair of conjugate, purely imaginary eigenvalues and C_2 be the subset of matrices with at least two such pairs, (here the eigenvalues are counted with multiplicity). From the M-Structure Lemma (R. Abrahams and J. Robin: <u>Transversal mappings and flows</u>. W. A. Benjamin N.Y. 1967 p.99), C_1 and C_2 are the union of closed submanifolds of $M(2k)$ of codimension respectively positive and greater than one.

Approximate then $B(\lambda)$ by a C^∞ matrix such that $B(\lambda_o \pm \rho)$ remain unchanged: For example by $(\lambda_o+\rho-\lambda)/2\rho\ B(\lambda_o-\rho)+(\lambda-\lambda_o+\rho)/2\rho\ B(\lambda_o+\rho)$. By choosing ρ small enough and r large enough (r may have any value smaller than $2\pi/\max k_j\beta$) the continuity of the eigenvalues imply that the class of the new map is unchanged. Apply the Thom transversality theorem (R. S. Stong: <u>Notes on Cobordism theory</u>. Princeton U. Press. Princeton N.J. 1968. Appendix) and perturb $B(\lambda)$ on $|\lambda-\lambda_o| \leq \rho$ to a path starting from $B(\lambda_o-\rho)$ and ending at $B(\lambda_o+\rho)$ which meets transversally C_1 at $\lambda_1 < \lambda_2 < \ldots < \lambda_\ell$ and avoids C_2. That is $B'(\lambda)$ has ℓ pairs of purely imaginary simple eigenvalues $\pm i\beta_j$ for $\lambda = \lambda_j$. Note that if $n_\pm$ is the number of $\alpha_j(\lambda_o \pm \rho)$ which are positive then $n_+ - n_-$

has the parity of the multiplicity of β. Also, if $n^j_\pm$ is the number of eigenvalues of $B'(\lambda_j \pm \varepsilon)$ with positive real part, then: $n_+ - n_- = n^\ell_+ - n^1_- = \sum_{j=1}^{\ell} n^j_+ - n^j_-$ (since $n^{j-1}_+ = n^j_-$).

Divide the interval $[\lambda_o - \rho, \lambda_o + \rho]$ into ℓ intervals each containing exactly one λ_j in its interior. Then the homotopy class of $I - \exp tB'(\lambda)$ on $S^1(\rho, r)$ will be the sum of the homotopy elements corresponding to the small rectangles around (λ_j, t_o), one can reduce its λ-size without changing the class, since in the process $I - \exp tB'(\lambda)$ will remain invertible. Hence reduce the size of the rectangle in order to repeat the diagonalization process: That is there is a continuous matrix $F(\lambda)$, which can be chosen with positive determinant, such that for $|\lambda - \lambda_j| < \rho_j$

$$B'(\lambda) = F(\lambda) \begin{bmatrix} D'_j(\lambda) & 0 \\ 0 & \tilde{D}_j(\lambda) \end{bmatrix} F(\lambda)^{-1}$$

where $D'_j(\lambda)$ is a 2×2 matrix with eigenvalues $\mu_j(\lambda) = \alpha_j(\lambda) \pm i\beta_j(\lambda)$ with $\alpha_j(\lambda) \neq 0$ for $\lambda \neq \lambda_j, \alpha_j(\lambda_o) = 0, \beta_j(\lambda_o) = \beta_j$ is close to $k_j\beta$, and $\tilde{D}_j(\lambda)$ has no purely imaginary eigenvalues.
Then, on the boundary of the small rectangle, $I - \exp tB'(\lambda)$ can be deformed to:

$$\begin{bmatrix} I - \exp tD'_j(\lambda) & 0 \\ 0 & I_{2k-2} \end{bmatrix}$$

deforming first $F(\lambda)$ to $F(\lambda_j)$ and by taking a path in $GL(2k)^+$ from $F(\lambda_j)$ to the identity, then one deforms $\tilde{D}_j(\lambda)$ to $\tilde{D}_j(\lambda_j)$

and $I-\exp t\tilde{D}_j(\lambda_j)$ to I_{2k-2}; (Note that the determinant of $I-\exp t\tilde{D}_j(\lambda)$ is positive since it is the product of terms of the form: $1 + \exp 2t\alpha - 2 \exp t\alpha \cos t\beta \geq (1 - \exp t\alpha)^2 > 0$, since the eigenvalues of $\tilde{D}_j(\lambda)$ have non-zero real part and the perturbation of $B(\lambda)$ can be made so small that these eigenvalues remain complex conjugate.).

Now there exists a matrix C with positive determinant such that $C^{-1} D'_j(\lambda_j)C$ is in canonical form: $\begin{bmatrix} 0 & \beta_j \\ -\beta_j & 0 \end{bmatrix}$.

So, taking a path from I_2 to C, one may assume that $D'_j(\lambda_j)$ is in the above form and one can further deform $D'_j(\lambda)$ to

$$D_j(\lambda) = \begin{bmatrix} \alpha_j(\lambda) & \beta_j(\lambda) \\ -\beta_j(\lambda) & \alpha_j(\lambda) \end{bmatrix},$$

using the homotopy $(1-\tau)D'_j(\lambda) + \tau D_j(\lambda)$: These matrices are close and the eigenvalues remain complex conjugate so that the trace of the deformation (2 Real part of the eigenvalue) remain equal to $2\alpha_j(\lambda)$.

$I-\exp tD_j(\lambda)$ has the form:

$$\begin{bmatrix} 1-\exp t\alpha_j(\lambda) \cos t\beta_j(\lambda) & -\exp t\alpha_j(\lambda) \sin t\beta_j(\lambda) \\ \exp t\alpha_j(\lambda) \sin t\beta_j(\lambda) & 1-\exp t\alpha_j(\lambda) \cos t\beta_j(\lambda) \end{bmatrix}$$

Deform $\beta_j(\lambda)$ to β_j by $\beta_j(\tau\lambda_j+(1-\tau)\lambda)$: If the determinant vanishes then: the sinus must be zero, the cosinus one, $\alpha_j(\lambda)=0$ so $\lambda = \lambda_j$ and $t\beta_j = 2\pi k_j$ that is $t = 2\pi k_j/\beta_j = t_j$ is close to t_o.

Then, using $1-\exp\{t^{(1-\tau)}\alpha_j(\lambda)\}\{\tau + (1-\tau)\cos 2\pi\, k_j t/t_j\}$ and $\exp\{t\alpha_j(\lambda)(1-\tau)\}\{\tau(t-t_j) + (1-\tau)\sin 2\pi\, k_j t/t_j\}$

one reduces the matrix to: $\begin{bmatrix} 1-\exp\,\alpha_j(\lambda) & -(t-t_j) \\ (t-t_j) & 1-\exp\,\alpha_j(\lambda) \end{bmatrix}$

since $t-t_j$ and $\sin 2\pi\, k_j t/t_j$ have the same sign for t close to t_j (one may have to reduce the t-size of the rectangle).

There are now four cases:

a) $\alpha_j(\lambda)(\lambda-\lambda_j) > 0$ for $\lambda \neq \lambda_j$ $(n_+^j - n_-^j = 1)$.
then $(1-\tau)(1-\exp\,\alpha_j(\lambda)) - \tau(\lambda-\lambda_j)$ is zero only at $\lambda = \lambda_j$, so one can deform the matrix to: $\begin{bmatrix} -(\lambda-\lambda_j) & -(t-t_j) \\ (t-t_j) & -(\lambda-\lambda_j) \end{bmatrix}$ which, via a rotation, is the generator α of $\pi_1(GL(2)^+)$.

b) $\alpha_j(\lambda)(\lambda-\lambda_j) < 0$ for $\lambda \neq \lambda_j$ $(n_+^j - n_-^j = -1)$.

A similar deformation will give: $\begin{bmatrix} \lambda-\lambda_j & -(t-t_j) \\ t-t_j & \lambda-\lambda_j \end{bmatrix}$ which is the homotopy inverse of α.

c) $\alpha_j(\lambda) > 0$ for $\lambda \neq \lambda_j$ $(n_+^j - n_-^j = 0)$.
$(1-\tau)(1-\exp\,\alpha_j(\lambda)) - \tau$ and $-(1-\tau)(t-t_j)$ will deform the matrix to $-I$ which is trivial.

d) $\alpha_j(\lambda) < 0$ for $\lambda \neq \lambda_j$ $(n_+^j - n_-^j = 0)$.
A similar deformation will give I.

So the class of $I-\exp tL(\lambda)$ is the suspension of $\sum_{j=1}^{\ell} (n_+^j - n_-^j)\alpha = (n_+ - n_-)\alpha$ which is non-trivial if and only if the multiplicity of β is odd.

Note: On page 93 it is claimed that if P is the set:

$$P = \{(x,\lambda,t) \; ; \; t > 0 \; / \; F(x,\lambda,t) = 0 \text{ and } f(x,\lambda) \neq 0\}$$

and if $S = \{(x,\lambda,t) \; / \; f(x,\lambda) = 0, \; t \geq 0\}$ then $\overline{P} - P \subset S$. Here the closure is taken with respect to $\mathbb{R}^{n+2}$.

In fact, if (x_n,λ_n,t_n) tends to (x_1,λ_1,t_1) which does not belong to P and if $t_1 > 0$, then the continuity of F implies that $F(x_1,\lambda_1,t_1) = 0$ and so one must have: $f(x_1,\lambda_1)=0$ On the other hand if $t_1 = 0$, let t' be any positive real number and let k_n be the smallest integer such that: $t_n k_n \geq t'$; that is $k_n = t'/t_n + \varepsilon_n$ with $0 \leq \varepsilon_n \leq 1$. Then $t_n k_n$ tends to t' as n goes to ∞ and $G(x_n,\lambda_n,t_n k_n) = x_n$ by periodicity (t_n is a period of $G(x_n,\lambda_n,t)$). So by continuity: $G(x_1,\lambda_1,t') = x_1$ for all positive t', that is $f(x_1,\lambda_1) = 0$.

This explains that in the proof of theorem 3.1 of chapter three one can choose the open set Ω_1 at a positive distance of the set $\{(x,\lambda,0)\}$.

BIBLIOGRAPHY

[0] J. C. Alexander and J. A. Yorke: Global bifurcation of periodic orbits (revised version). U. of Maryland Technical report 1974.

[1] M. S. Berger: Bifurcation theory and the type numbers of M. Morse, Procedings Nat. Acad. of Sciences, Vol. 69, No. 7, July 1972, pp. 1737-1738.

[2] M. S. Berger and M. S. Berger: _Perspectives in nonlinearity_, Benjamin, New York, 1968.

[3] N. A. Bobylev and M. A. Krasnosel'skii: Operators with continua of fixed points, Dokl. Akad. Nauk. SSSR, Tom 205 (1972), No. 5. Translated in Soviet Math. Dokl. Vol. 13 (1972), No. 4, 1045-1049.

[4] R. Böhme: Die Lösung der Verzweingungsgleichungen für nichtlineare eigenwertprobleme, Math. Z. 127, 105-126 (1972).

[5] M. G. Crandall and P. H. Rabinowitz: Bifurcation for simple eigenvalues, Journal of Functional Analysis, 8, 321-340, (1971).

[6] J. Cronin: _Fixed points and topological degree in nonlinear analysis_, A.M.S., 1964 (Providence).

[7] __________: Using Leray-Schauder degree, Journal of Math. Analysis and Applications, 25, (1969), 414-424.

[8] __________: Upper and lower bounds for the number of solutions of nonlinear equations, Proc. Symposium Pure Math., Vol. 18, A.M.S., 1970.

[9] E. N. Dancer: Bifurcation theory in real Banach space, Proc. London Math. Soc. (3), 23 (1971), 699-734.

[10] ____________: Bifurcation theory for analytic operators, Proc. London Math. Soc. (3), 26 (1973) 359-384.

[11] A. Friedman: Partial differential equations, Holt Rinehart and Winston, (1969).

[12] W. Fulton: Algebraic curves, Benjamin (1969).

[13] K. Geba: Algebraic methods in the theory of compact vector fields in Banach spaces, Fundamenta Math. 54 (1964), 168-209.

[14] S. Goldberg: Unbounded linear operators, McGraw-Hill (1966).

[15] I. C. Gokhberg and M. G. Krein: Fundamental theorems on deficiency numbers, root numbers and indices of linear operators. Uspehi Mat. Nauk. 12 (1957), 43-118 translated in: Amer. Math. Soc. Transl. (2), 13 (1960), 185-264.

[16] A. Granas: Theory of compact vector fields. Rosprawy Matematyczne (1962).

[17] _________: Topics in infinite dimensional topology, Séminaire sur les équations aux derivées partielles Collège de France (1969-1970).

[18] R. Gunning and H. Rossi: Analytic functions of several complex variables, Prentice Hall (1965).

[19] H. Hopf: Abbildung der dreidimensionalen Sphäre auf die Kugelfläche, Math. Annalen 104 (1931), 637-665.

[20] T. Kato: Perturbation theory for linear operators. Springer Verlag, Berlin (1966).

[21] K. Kirchgässner: Multiple eigenvalue bifurcation for holomorphic mappings, Contributions to nonlinear functional analysis, 69-100. Academic Press (1971).

[22] M. A. Krasnosel'skii: <u>Topological methods in the theory of non-linear integral equations</u>, Macmillan, New York (1964).

[22]' ______________: <u>Plane vector fields</u>. Academic Press (1966).

[23] V. J. Naumann: Lusternik-Schnirelman Theorie und nicht-lineare eigenwertprobleme. Mathematische Nachrichten, 53 (1972), 303-336.

[24] L Nirenberg: An application of generalized degree to a class of nonlinear problems, Troisième Colloque d'analyse fonctionelle Liège, Sept. 1970.

[25] ______________: <u>Topics in nonlinear functional analysis</u>, NYU Lecture Notes, 1973-74.

[26] P. H. Rabinowitz: A global theorem for nonlinear eigen-value problems and applications, Contributions to nonlinear functional analysis, 11-35. Academic Press (1971).

[27] ______________: Some aspects of nonlinear eigenvalue problems, Rocky Mountain Journal of Math., Vol.3, No. 2, Spring 1973, 161-202.

[28] E. H. Rothe: A relation between the type numbers of a critical point and the index of the corresponding field of gradient operators, Math. Nachrichten 4, 1950-51, 12-27.

[29] ______________: Morse theory in Hilbert space, Rocky Mountain Journal of Math. Vol. 3, No. 2, Spring 1973, 251-274.

[30] D. Sather: Branching of solutions of nonlinear equations, Rocky Mountain Journal of Math. Vol. 3, No. 2, Spring 1973, 203-250.

[31] M. Schechter: Principles of functional analysis. Academic Press (1971).

[32] J. T. Schwartz: Nonlinear functional analysis, Gordon and Breach (1969).

[33] E. Spanier: Duality and S-theory, Bull. Amer. Math. Soc. 62 (1956), 194-203.

[34] ________: Algebraic topology McGraw Hill (1966).

[35] H. Toda: Composition methods in homotopy groups of spheres, Annals of Math. Studies, No. 49, Princeton U., 1962.

[36] A. E. Taylor: Introduction to functional analysis. New York, Wiley and Sons (1958).

[37] ________: Theorems on ascent, descent, nullity and defect of linear operators, Math. Annalen, 163, 18-49 (1966).

[38] M. M. Vainberg: Variational methods for the study of non-linear operators, Holden day Series in Math. Physics (1964).

C.I.M.A.S.
U.N.A.M.
MEXICO.